T0256239

Creative Economy and Sustainable Development

The creative economy is one of the world's most dynamic sectors. Drawing upon the author's work on empowerment and sustainability, this book focuses on India's indigenous, rural, traditional handicraft-based creative and cultural industries (CCIs) and the role they can play in the country's creative economy.

The book combines a comprehensive assessment of the region's deeply rooted cultural and creative resources with practical cases of self-sufficient creative skills and knowledge-based entrepreneurship across the Indian handicrafts sector. The author illuminates how sustainability, resilience, and collective well-being, along with unique regional characteristics, are converging towards generating an independent creative and cultural economy that does not depend on global brands and businesses alone. The disconnect between associated policies, practice, and academic work is addressed by contextualizing the case studies in terms of modern economic theory and practice, relevant administrative policies of South Asia, and recognition of the role of culture in achieving the sustainable development goals.

This concise yet comprehensive book provides an insightful and holistic understanding of India's handicrafts economy which will be valuable reading for researchers and reflective practitioners.

Madhura Dutta, PhD is a development sector professional, researcher, and writer with more than two decades of national and international experience in culture-based sustainable development programs.

Routledge Focus on the Global Creative Economy
Series Editor: Aleksandar Brkić, *Goldsmiths,*
University of London, UK

This innovative Shortform book series aims to provoke and inspire new ways of thinking, new interpretations, emerging research, and insights from different fields. In rethinking the relationship of creative economies and societies beyond the traditional frameworks, the series is intentionally inclusive. Featuring diverse voices from around the world, books in the series bridge scholarship and practice across arts and cultural management, the creative industries, and the global creative economy.

Creative Work Beyond Precarity
Learning to Work Together
Tim Butcher

Youth Culture and the Music Industry in Contemporary Cambodia
Questioning Tradition
Darathtey Din

Global Crisis and the Creative Industries
Analysing the Impact of the Covid-19 Pandemic
Ryan Daniel

Creative Economy and Sustainable Development
The Context of Indian Handicrafts
Madhura Dutta

For more information about this series, please visit: www.routledge.com/ Routledge-Focus-on-the-Global-Creative-Economy/book-series/RFGCE

Creative Economy and Sustainable Development

The Context of Indian Handicrafts

Madhura Dutta

Routledge
Taylor & Francis Group

LONDON AND NEW YORK

First published 2024
by Routledge
4 Park Square, Milton Park, Abingdon, Oxon OX14 4RN

and by Routledge
605 Third Avenue, New York, NY 10158

Routledge is an imprint of the Taylor & Francis Group, an informa business

© 2024 Madhura Dutta

The right of Madhura Dutta to be identified as author of this work has been asserted in accordance with sections 77 and 78 of the Copyright, Designs and Patents Act 1988.

British Library Cataloguing-in-Publication Data
A catalogue record for this book is available from the British Library

ISBN: 9781032363448 (hbk)
ISBN: 9781032363462 (pbk)
ISBN: 9781003331476 (ebk)

DOI: 10.4324/9781003331476

Typeset in Times New Roman
by Newgen Publishing UK

Contents

Preface

The term *creative industry* covers a range of art form-based businesses—architecture, crafts and design, performing arts, visual arts, music, film, literature, art journalism, illustrations and comics, animations, multidisciplinary art, and even circus art. Television, radio, and multi-media are all part of this industry. Together, all these creative industries form the *Creative Economy* of a country (Howkins 2001).

The creative economy is one of the most dynamic and expansive sectors in today's world economy, involving and employing diverse actors. It has various associated industries, such as tourism and education, but the nature, potential, robustness, and impact of the creative economy of a country depend primarily on its existing cultural, human, and natural resources, its cultural and political history, and its demography. As these factors vary substantially from country to country, creative economy is often understood and approached very differently across the world.

India has a rich creative industry, well known for its cinema, fine arts, classical music and dance, literature, and multidisciplinary art. But it also has an enormous wealth of indigenous, rural artform-based traditional *cultural* industries which are less well-known and more informal in nature. Most of these indigenous art forms have very interesting cultural origins and history, with strong links to ancient trade routes and traditions. However, the disparity between the two is alarming, with the traditional rural art forms being almost insignificant in the national creative economy, even though they bear a strong influence on the more commercially successful urban creative industries in many ways.

Although many actors and stakeholders have been actively working in the traditional cultural industries of rural India for decades, the sector has remained niche and its output has largely been perceived in the national economic arena as the product of cultural heritage and leisurely activities, rather than economic activity. However, in the post-pandemic world, the appreciation of traditional cultural industries, especially based on the intangible cultural heritage of India, has increased significantly among sectoral stakeholders, if not in the public consciousness. This change is driven by the realization that rural local economies based on local resources and capabilities are more resilient and sustainable, and can recuperate faster in the face of a collective crisis. Our aim in this book is to highlight successful case studies that focus on this aspect of traditional cultural industries of India, and the potential of these models to create a vibrant creative economy for the modern age.

The creative economy and cultural heritage

Creative economies based on traditional cultural heritage is not, by any means, a new concept. It has been a topic of discourse since the 1980s, both at the levels of national and regional approaches to re-strategizing economic growth, as well as at international policy levels. In the 1980s UNESCO identified culture, as defined by the arts and the heritage of a nation, as a driver of sustainable development contributing directly to socio-economic development and human well-being. However, consciousness about culture as a driver of socio-economic development was slow to evolve over the years, owing to the diverse and expansive nature of the sector and the lack of an inclusive definition. Although several countries had independently started focusing on their respective creative industries, these were mostly national initiatives to assess internal wealth creation activities and their subsequent impact on the national and local economy.

Since the 1990s, global economic changes have accelerated the growth of creative economies by enabling the quick flow of creative ideas, products, and services. This growth has been supported by advances in communication technologies, changing ways of business organization, international exchange and market linkages, the increasing tradability of knowledge and services, and globalization, leading to the worldwide integration of cultures, markets, and consumers. The potential of traditional cultural heritage as a contributor to economic growth has undergone a reassessment during this time. Before the digital era, culture was largely confined to the domain of national pride and 'maintaining tradition.' In the digital world, where technology has made the world a more closely connected interdependent place, the semantics of creative economy and cultural industries have shifted towards strategies for reconstructing culture as skill-based businesses, enhancing marketability in export markets, and effective enterprise dynamics.

From an academic perspective, this global shift towards creative economies, comprising creative and cultural industries, is better understood through the framework of evolutionary economics, rather than mainstream neoclassical economics. The evolutionary economics approach discards the rational choice theory of traditional economics and proposes that economic processes evolve and are determined both by individuals and society as a whole, with psychological factors being key drivers of the economy. Evolutionary economists believe that the economy is dynamic, constantly changing, and chaotic, rather than always tending towards a state of equilibrium (Nelson and Winter 1982; Saviotti 1996). Hence, the more pertinent analytical foundation proposed for creative industries is that of a market economy driven by the nature of markets that shape this industry, and not the neoclassical industrial economy (Potts et al. 2008; Potts 2011).

The United Nations Conference on Trade and Development (UNCTAD) has been instrumental in promoting and analyzing creative economies all over the world through its Creative Economy Program. In the first comprehensive report published by the UN on this topic, UNCTAD and UNDP (2008) described the creative economy as 'an evolving concept based on creative assets potentially generating economic growth and development'. On its website, UNCTAD now characterizes creative economy as one that builds on the interplay between human creativity, intellectual property, knowledge, and technology.[1] It is now clear that the creative economy has the potential to not only strengthen the expansion of global value chains, increase digital adoption among creative small and medium-sized enterprises (SMEs), fuel the export of cultural goods and creative services, and foster ownership through local engagement, but also contribute to the overarching goal of sustainable development (Sirivunnabood and Alegre 2021). Looking at the global South and specifically the case of India, Limaye (2020) notes that the cultural and creative industry has the potential to address the three pillars of sustainable development—economic, social, and environmental—by generating income and employment opportunities leading to national economic benefits, foreign trade, and investments, improving social cohesion, and building eco-friendly businesses that depend primarily on intangible resources such as traditional knowledge and skills.

To formally bind the different nations in creating and implementing policies to protect, safeguard, enhance, and promote the world's cultural heritage and cultural diversity for achieving people-centered sustainable development, UNESCO established two important Conventions, in 2003 and 2005.

The 2003 Convention for the Safeguarding of the Intangible Cultural Heritage was adopted on October 17, 2003, at the 32nd session of the UNESCO General Conference. The Convention not only defined Intangible Cultural Heritage (ICH) and the domains manifesting ICH but also laid out the importance of ensuring respect and mutual appreciation for ICH assets. These assets constitute the cultural capital of a country which directly addresses issues of food security, health, education, sustainable use of natural resources, women empowerment, etc. The aim of this Convention is to ensure the viability of intangible cultural heritage, its identification, documentation, preservation, protection, promotion, enhancement, and transmission, particularly through formal and non-formal education, as well as the revitalization of the various aspects of such heritage with the support of international cooperation and assistance. A total of 180 countries have ratified this Convention. This Convention developed, for the first time, a common and shared understanding of our living heritage across the world.

Furthering the recognition of culture as a component of multisectoral and holistic development, the 2005 Convention on the Protection and Promotion of the Diversity of Cultural Expressions was adopted on October 20, 2005, at the 33rd session of the UNESCO General Conference. The 2005 Convention reaffirmed the values of the Universal Declaration of Human Rights and the UNESCO Universal Declaration on Cultural Diversity. This

Convention was a milestone in the area of international cultural policy, as it laid out both the cultural and economic aspects of ICH and the need for effective and equitable governance of cultural assets. Thus, it provides guidelines for the implementation of policies and measures that support the creation, production, distribution of, and access to cultural goods and services, and focuses on the importance of a creative economy.

As frameworks and indicators for measuring the impact and economic outputs of creative economies evolved, it was realized that the sector not only contributes to economic growth but improves social empowerment, inclusion, and emotional well-being as well. Thus, in 2015, culture was recognized as a cross-cutting area for achieving the 2030 agenda for Sustainable Development Goals (SDGs), based on various United Nations resolutions on culture and development. All the United Nations Member States adopted the agenda with the objective of sharing common goals and achievement targets for peace and prosperity for the people and the planet for the present and the future. A set of seventeen SDGs were laid out, consolidating all the urgent action areas for both developing and developed countries relating to poverty alleviation, ending deprivations and marginalization, improving universal access to health and education, reducing inequality, along with addressing climate change and unsustainable practices and lifestyles, and strengthening conservation of nature. This led to a more consolidated global approach to the creative economy. As evidence-building is a crucial part of tracking the 2030 SDG agenda, there was greater focus on data, case studies, experience sharing, and policy dialogues.

A 2015 study commissioned by CISAC (International Confederation of Societies of Authors and Composers) and carried out by EY estimated that the Asia-Pacific region accounts for 43% of jobs in the cultural and creative industries worldwide, with visual arts, books, and music industries as the main employers (EY 2015). The 2018 Global Report on the 2005 Convention (UNESCO 2017) estimated the global revenue of creative industries to be 2250 billion USD, with the sector employing 30 million people, of whom 20% were younger than 30 years, and 45% were women in cultural occupations.

On the occasion of the World Day for Cultural Diversity for Dialogue and Development in 2019, the United Nations organized a high-level event on Culture and Sustainable Development where it reaffirmed the transformative power of culture for social inclusion, resilience, and sustainable development, stating that[2]

> Culture is also an essential component of human development, representing a source of identity, innovation and creativity for all, it provides sustainable solutions to local and global challenges.

Recognizing the potential of the sector, the United Nations declared 2021 as the International Year for Creative Economy for Sustainable Development, in the 74th UN General Assembly (UNESCO 2021).

The UNESCO World Conference on Cultural Policies and Sustainable Development—Mondiacult 2022 (UNESCO Mondiacult), hosted in September 2022 by the Government of Mexico, was the most recent and significant step towards fully integrating culture and sustainable development. It was convened by UNESCO 40 years after the first Mondiacult World Conference on Cultural Policies held in Mexico City (Mexico) in 1982, and 24 years after the UNESCO World Conference on Cultural Policies for Development held in Stockholm (Sweden) in 1998. Here, a new report was issued entitled 'Stormy Times. Nature and humans: Cultural courage for change.' It was the largest world conference devoted to culture in the last 40 years, which brought together nearly 2,600 participants over three days in Mexico City. At the end of the three-day conference, 150 states unanimously adopted a Declaration for Culture, where culture was affirmed as a 'global public good.' It was agreed upon that culture should be incorporated in public policies focusing on 'social and economic rights of artists, artistic freedom, the right of indigenous communities to safeguard and transmit their ancestral knowledge, and the protection and promotion of cultural and natural heritage.'

Why this book?

In the backdrop of these ongoing international efforts to link culture with economic growth and sustainable development, our aim is to undertake an assessment of the traditional cultural heritage of India, specifically its traditional handicrafts industry, and how its potential to form the basis of a sustainable and equitable creative economy can be fully realized. One of the first significant international symposiums on creative cultural industries in India was organized in 2005, where 28 countries participated at the Senior Experts Symposium on Asia-Pacific Creative Communities held in Jodhpur. The symposium was convened by UNESCO, UNIDO, WIPO, ADB, and the World Bank with the support of The Indian National Trust for Art and Cultural Heritage (INTACH). A new world vision was discussed on how these creative industries can be strong contributors towards the region's economic and social development, and hence need to be integrated with national policies and programs and strong public-private partnerships. It was put forward in the consensus report that creativity and creative communities are among the most important resources at hand in the developing world. The meeting adopted the 'Jodhpur Consensus' (UNESCO 2008), which stated that

> the region's development challenge requires cultural industries, and the creative communities in which they are found, to be fully recognized as a source of capital assets for economic, social and cultural development. We must use these assets to empower these communities, alleviate poverty, and sustain and strengthen our diverse cultures.

To understand the true potential of these creative cultural industries and the challenges facing them, the first two chapters of the book review the history of the creative economy in India and contextualize it in terms of modern economic theory and practice. In Chapter 1, we take a closer look at the history and evolution of creative industries in India, centered around traditional handicrafts (including handloom), which goes back to its ancient civilizations and is a testament to the extremely rich cultural wealth of the country which has impacted global trade and commerce in significant ways. In Chapter 2, we briefly discuss the theoretical foundations of people-centric economic approaches, both from theoretical as well as policy viewpoints, and how they relate to the creative economy.

The remaining chapters focus on case studies from diverse parts of India illustrating the practical and real-life socio-economic impact of traditional handicraft-based creative industries and discuss the underlying models that have the potential to invigorate the sector further.

While reading this book, it will be useful to keep the bigger picture in mind. It is perhaps obvious that the creative economy of a country can flourish only at times of political and financial stability. At the same time, it is also important to note that administrative policy, whether explicitly or implicitly, plays a key role as well. At the height of its glory in the pre-industrial era, the creative economy of India was enabled by the administration, financial systems, patronage, and religions of the time. In the absence of many of these factors in the modern world, the revitalization of trade and commerce in traditional creative industries, such as handicrafts, requires well-reasoned policy mechanisms. Although considerable effort has been made in this direction, substantial gaps still remain.

We hope that the case studies presented in the book can serve to highlight some of these gaps and serve as models that can be adapted or scaled, with appropriate policy support, to build a renewed, modern, and sustainable creative economy that leverages the huge traditional cultural wealth of India, and indeed Asia.

Notes

1 https://unctad.org/topic/trade-analysis/creative-economy-programme (accessed May 2023).
2 www.un.org/pga/73/event/culture-and-sustainable-development/ (accessed May 2023).

References

EY. 2015. Cultural Times: The First Global Map of Cultural and Creative Industries. UNESCO, Paris. https://unesdoc.unesco.org/ark:/48223/pf0000235710.

Howkins, John. 2001. *The Creative Economy: How People Make Money from Ideas*. Penguin UK.

Limaye, Adwait. 2020. "Cultural and Creative Industry (CCI) Policy for Sustainable Development, a Case of India." In *Proceedings of the International Conference on Sustainable Development 2020*. https://ic-sd.org/2020/11/21/proceedings-from-icsd-2020/.

Nelson, Richard R, and Sidney G. Winter. 1982. *An Evolutionary Theory of Economic Change*. Harvard University Press, Cambridge, Massachusetts; London, England.

Potts, Jason. 2011. *Creative Industries and Economic Evolution*. Edward Elgar Publishing Limited, Cheltenham, UK.

Potts, Jason, Stuart Cunningham, John Hartley, and Paul Ormerod. 2008. "Social Network Markets: A New Definition of the Creative Industries." *Journal of Cultural Economics* 32: 167–85.

Saviotti, Pier Paolo. 1996. *Technological Evolution, Variety and the Economy*. Edward Elgar Publishing Limited, Cheltenham, UK.

Sirivunnabood, Pitchaya, and Herlyn Gail A Alegre. 2021. "Supporting the Creative Economy for Sustainable Development in Southeast Asia." Asian Development Bank Institute. www.asiapathways-adbi.org/2021/09/supporting-creative-economy-sustainable-development-southeast-asia/.

UNCTAD, and UNDP. 2008. "Creative Economy Report." United Nations, Geneva. https://unctad.org/system/files/official-document/ditc20082cer_en.pdf.

UNESCO. 2008. "The Jodhpur Initiatives: A Strategy for the 21st Century." UNESCO. https://unesdoc.unesco.org/ark:/48223/pf0000179570.

———. 2017. "Re/shaping Cultural Policies: Advancing Creativity for Development." https://uis.unesco.org/sites/default/files/documents/reshaping-cultural-policies-2018-en.pdf; UNESCO Paris.

———. 2021. "International Year of Creative Economy for Sustainable Development." UNESCO. www.unesco.org/en/articles/international-year-creative-economy-sustainable-development.

Abbreviations

AAJ	Artisans' Alliance of Jawaja
ADB	Asian Development Bank
AIACA	All India Artisans and Craftworkers Welfare Association
AIHB	All India Handicrafts Board
CFC	Common Facility Center
CLRI	Central Leather Research Institute
EPCH	Export Promotion Council for Handicrafts
GI	Geographical Indication
ILO	International Labour Organization
INTACH	Indian National Trust for Art and Cultural Heritage
JLA	Jawaja Leather Association
KVIC	Khadi and Village Industries Commission
MSME&T	Department of Micro, Small and Medium Enterprises and Textiles
NID	National Institute of Design
NGO	Non Government Organization
OECD	Organisation for Economic Co-operation and Development
OPHI	Oxford Poverty and Human Development Initiative
UNDP	United Nations Development Programme
UNESCO	United Nations Educational, Scientific and Cultural Organization
UNIDO	United Nations Industrial Development Organization
UNWTO	World Tourism Organization

WCED	World Commission on Environment and Development (WCED)
WIPO	World Intellectual Property Organization

1 History and evolution of creative industries in India

The pre-Islamic era

India has a rich history of traditional creative industries that can be traced back to the Indus Valley Civilization (3300–1300 BCE), which produced sculptures, seals, pottery, and jewelry from local materials, such as terracotta, metal, and stone. Indications of cotton weaving have also been found. Evidence of contemporary trade with other civilizations indicates that the Indus Valley Civilization had access to a wealth of unique raw materials and produced marketable items that were in demand in other parts of the world.

During the Vedic age (1500–600 BCE), spanning the later part of the Bronze Age and the early Iron Age, the arts of pottery, sculpture (metal, stone, and terracotta), jewelry, weaving, etc., developed further. There was a growing metal industry, along with leather craft and carpentry, resulting from the diffusion of new knowledge of techniques brought in by traders. These two early periods are usually associated with the type of pottery excavated from archaeological sites, and are commonly referred to as Black and Red Ware culture and Painted Grey Ware culture (Mallory and Adams 1997).

Through the Iron Age, Mauryan and Gupta periods, and Mughal rule, art and crafts flourished significantly with advanced craftsmanship, knowledge, and technology, and were regularly traded in domestic and foreign markets.

The late Vedic period (600–300 BCE) saw the emergence of large city-states or *mahajanapadas* in Northern India and was followed by the rise of the Mauryan Empire (322–185 BCE). Archaeologically,

DOI: 10.4324/9781003331476-1

this period is linked to the Northern Black Polished Ware culture. The Mauryan Empire was marked by exceptional creativity in art, architecture, and inscriptions. Under the Mauryas, internal and external trade expanded across South Asia, with India exporting silk goods and textiles, spices, and exotic foods.

After a few centuries of relative fragmentation, the Gupta Empire (319–467 CE) brought back political and administrative stability to a large part of India. Often referred to as the Golden Age in Indian history, this was a time of monumental achievements in the fields of architecture, sculpting, paintings, craft, weaving, academics, literature, poetry, music, science, and technology (Harle 1994; Agrawal 1989). The period was known for its exceptional skills in jewelry work with gemstones, gold, and silver. Silk fabric, muslin, linen, woollen, and cotton textiles were also in great demand in foreign countries. Strong trading relationships along with investment in art and culture made India an important cultural center. International trade, particularly with the contemporary Roman Empire, was an important source of revenue, and its disruption, likely due to invasion by the Alchon Huns from Central Asia, contributed to the eventual decline of the Gupta Empire.

The Islamic era

Following the disintegration of the Gupta Empire, the Indian subcontinent was ruled by a succession of smaller nation states and kingdoms, before being united again under the Delhi Sultanate. The Delhi Sultanate persecuted both Hindu and Buddhist elites, desecrated religious structures, destroyed centers of learning, and generally disrupted classical Indian culture to a serious extent. At the same time, it probably protected the subcontinent from Mongol invasion and established new avenues of trade and cultural relations with the growing Islamic world, particularly Persia (Asher and Talbot 2006). Some historians believe that the technology of the iconic Charka or spinning wheel was imported from Persia during this period (Pacey 1993). Domestic and international trade in India's creative economy continued in this period, through both the Delhi Sultanate in the north and the Vijayanagara Empire in the south (Nilakanta Sastri 1955).

The Delhi Sultanate was eventually defeated and replaced by the Mughal Empire, although the breakaway Bengal Sultanate remained an important economic power for a considerable period before being absorbed into the Mughal Empire. The Mughal era in India lasted from the early 16th to mid-18th century and left a significant mark on the nation's history and culture. This period saw a proliferation of architecture, paintings, and diverse cultural contributions that amalgamated the styles of the Islamic world with Indian traditions. Indo Persian paintings, calligraphy, frescoes, intricate marble and stone carvings, the use of precious stones in architecture, and the use of ornamented tiles were some of the iconic elements of the creative industry during this time, all of which have had a strong influence on the evolution of art and craft in India and South Asia. International trade continued unabated, with the cotton textile industry based in Bengal contributing significantly.

From the 2nd century BCE to the middle of the 15th century, international trade and exchange with the east and the west were largely facilitated by the Silk Road or Silk Route network. Extending more than 6,500 km, this largely overland network connected East and South-East Asia, the Indian subcontinent, Central Asia, the Middle East, Europe, and East Africa (Boulnois, Mayhew, and Sheng 2004). People travelled in camel and horse caravans, as well as on maritime routes to explore trade as well as people-to-people relations. Trade not only ensured business but also developed avenues for the exchange and promotion of art, religion, folk industries, cultural awareness, and solidarity. The Silk Route's contribution to history is not limited to the history of international trade but also extends to the proliferation of ideas, art, and science between the countries involved. Four major corridors through the passes of high mountain ranges of the Himalayas and the Great table land of Tibet, and the river valleys of the Ganges, linked India to the ancient Silk Route.

The importance of the Silk Route declined with the rise of the Ottoman Empire in the 15th century CE, which disrupted trade relations between east and the west. This led to the eventual rise of alternative maritime trade routes between Europe and the East, indirectly ushering in the age of European colonialism.

The success of the long-lasting Mughal Empire can be ascribed to the effective administration and governance of the rulers that were based on the principles of inclusion of the people they conquered in their administration and military. In the later period, as the Mughal rulers started becoming autocratic and intolerant, widespread resentment among their subjects contributed to the breakdown of the empire. Internal rivalries and dynastic wars made it easy for Nader Shah, who seized power in Persia after the Safavid dynasty declined, to invade north India in 1739 and sack Delhi, vastly depleting the Mughal treasury and effectively weakening its control over most of India. Although the Maratha Empire, the Nawabs (rulers) of Bengal, and the Kingdom of Mysore took control of most of India after 1748, the British eventually gained effective supremacy over the course of the 18th century. The British East India Company, established in 1600, was initially interested in trade with the Mughal Empire but seeing the opportunity also defeated the already weakened empire. In 1757 the British defeated the Nawab of Bengal and French forces at the Battle of Palashi (Plassey). The East India Company thereafter took political control of much of the Indian subcontinent and exiled the last Mughal emperor during the Indian mutiny of 1857–59, following which the British crown took formal political control of most of India.

Colonial rule

Throughout the pre-Islamic and Islamic eras, from before the 1st century BCE to the 15th century CE, India remained one of the largest economies of the world, along with China (Maddison 2003). While the Industrial Revolution led to independent economic growth in the western world, the centuries old and internationally renowned creative industry of India was severely exploited and systematically decimated during colonial rule.

Before the British monopolized trade in Bengal and India, other European traders had their stints in the business of Bengal textiles. The Portuguese, Armenian, Danish, and French traders were also attracted to the wealth of cultural and natural resources, especially of eastern India, along the banks of the river Hooghly, and left their mark through their trading activities. These European traders radically transformed the textile trade from something local to global, increasing it in both scope and scale, with increased demand for Bengal textiles in the European markets, especially Bengal muslin and silk. Even printed cotton textiles exported from India became popular in England and other European countries from the 1680s. Although several models of private enterprises and businesses emerged involving rural weavers, weaver supervisors, intermediaries, and local traders, as well as large corporations and multinational trading enterprises, internal trade policies and competition between the European countries led to changing power dynamics of these foreign traders, finally leading to the monopolization, invasion, and colonization of India by the British.

Under colonial rule, cheaper factory systems of production requiring simpler management proliferated, and the hiring of skilled weavers as wage laborers disrupted the more self-sufficient village economies based on barter systems. During the 1760s, the emergence of the Industrial Revolution catalyzed industrial growth in England, leading to technologically advanced production systems, including that of textiles which could compete much better in terms of longevity, strength, and price in the world market. This led to a complete decline of demand for hand-woven Indian textiles, leaving thousands of weavers unemployed. Eventually, India became a supplier of raw materials such as cotton and indigo to England, and an importer of finished goods from England. This led to the complete destruction of the native creative industries of the country.

During the struggle for independence in the first half of the 20th century, efforts were made by national leaders, under the aegis of Mahatma Gandhi, to protect the traditional creative industries of India, which were mainly home-based or 'cottage' industries, by focusing on strengthening decentralized economic activities. After independence, Kamaladevi Chattopadhyay along with other visionaries pioneered the development of handicraft-based rural industries in national planning, recognizing the importance of artisanal cultures and livelihoods. However, the traditional cottage industries, of which a large fraction were local creative industries, underwent further disintegration in the post independence era with the acceleration of industrialization in India.

The post independence era

India is a predominantly rural country. As per the 2011 Census, 68.8% of the country's population and 72.4% of its workforce resided in rural areas. A discussion paper published by the NITI Aayog (Chand, Srivastava, and Singh 2017) projects that even with increasing urbanization, more than half the population will be rural by 2050. The rural economy constitutes 46% of the national income, with agriculture being the prime driver. The transition to more productive non-farm sectors is considered an important source of economic growth and transformation going forward.

In the pre-independence period, during British rule, rural resources were exploited by the colonizers for their own profit. But after independence, rural development was adopted in a planned manner through various programs and schemes. Although agriculture was the focus of the independent government, along with social upliftment and welfare (education, sanitation, infrastructure, upliftment of women and other disadvantaged groups, public health and hygiene, etc.), concepts of rural reconstruction and self-contained Indian villages also took root in government planning. The Gandhian approach of self-sufficient villages, through the promotion of village industries based on local resources and strengthening of traditional rural skills, led to the establishment of the landmark khadi and village industries program to support sustainable self-reliant village economies, among others.

The traditional cottage industries of rural India had historically been self-sufficient industries with full-time occupations that were not necessarily linked to agricultural activities. They included handicraft-based industries such as weaving, goldsmithing, carpentry, building-craft, pottery, etc., which were primarily linked to utilitarian and ritual products. These traditional industries, forming a major pillar of the pre-industrial economy, have been defined by the ILO as those industries that are characterized by the absence of power-driven machinery and the use of skills of craftspersons. In these industries, production is carried out in the place of residence of the artisans by the artisan family members, with almost no hired labor. The raw materials used are procured from local sources and the markets are also mostly local (Koga 1968). Highly skilled artisans, who constitute the cohort of master artisans, typically design and produce handicraft objects for local traders. The master artisans in turn have artisans under their supervision, who constitute their workforce and carry out production against wages.

The serious decline of these small-scale rural cottage industries started with the advent of modern industries and the threat that they posed through market competition. The establishment and proliferation of modern industries happened in India during the 1940s, 1950s, and especially post independence.

It is interesting to note that small-scale industries were present in many other advanced industrialized countries as well during the course of their economic development, but these were modern small-scale industries subsumed under their industrialized economic structures and not traditional industries as was the case in India. Although the traditional small-scale cottage industries persisted in India during industrialization after independence, they underwent a shift from a purely traditional and self-sufficient nature. For example, instead of local raw materials, the traditional industries started using raw materials produced by modern large-scale industries or began manufacturing semi-finished goods which then served as input for larger industries. With time, owing to westernization along with industrialization in India, traditional industries were supplanted by modern industries. For example, in Moradabad, Mirzapur, Rewari, and Jagadhri, there used to be household-based traditional industries manufacturing metal utensils, which were mechanized after the 1920s to become modern industries (Koga 1968). A 1963 survey of small-scale industries in Mysore shows that traditional silk weaving by handlooms was converted into weaving by powerlooms (Kapur 1963).

Rapid urbanization started in India following economic liberalization in the 1990s. The private sector emerged as a proponent of industrialization and the development of modern urban spaces. This resulted in the creation of jobs and labor work, drawing people to cities and towns from rural areas, resulting in the growth of urban India. Industrialization became synonymous with modernity, attracting rural populations to migrate to cities with the expectation of a better standard of living in the newly emerging non-agricultural economy. Industrial trade and commerce led to the emergence of modern marketing and exchange methods that held the promise of better returns. Urban acculturation, characterized by western notions of modernity, created new aspirations of growth and cultural and social changes.

This period also witnessed significant changes in people's behavioral and consumption patterns, increasing demand for non-farm industrial goods produced in factories. Urban infrastructure, technology, and availability of casual labor from the rural hinterlands created opportunities for manufacturing goods at scale and with reduced per capita cost. Such industrial developments became the mainstay of the modern Indian economy, contributing to increased national income. Urban centers thus symbolized advancement in technology, infrastructure, transportation, communication, modern education, medical facilities, and modern housing. Industrialization and steady urbanization led to major rural-urban migration with expectations of industrial jobs and better earnings in cities for the rural people. This increased pressure on the cities, which were unable to develop sufficient civic amenities, leading to challenges of overpopulation and unplanned urban growth. A large number of rural migrants with low-paid jobs were forced to live under deprivation. Due to the migration of the rural population to cities, an urban informal sector emerged which supplied cheap labor in response to the demands of the formal employment sector. The competitive advantage and negotiation power of these migrant laborers remained poor as they were mostly illiterate, not formally educated or trained, and caught between capitalist and feudal cultures.

In a sense, globalization had already started with the advent of the Silk Route and reached a new scale during the age of discovery starting in the 15th century, when explorers found new sea routes, lands, and people. However, it brought unprecedented changes in every aspect of life only with the introduction of information technology—the World Wide Web and the Internet—which revolutionized the world. The global era of interconnectedness, interdependency, and exchange led to major changes in economic, social, and cultural environments worldwide. As the world came closer, with the sense of a global community, it also made regional and local socio-cultural identities increasingly significant in terms of their values and ideologies, context and content, and the need for equality, fraternity, tolerance, and appreciation for other cultures. New opportunities for education, trade, policies, governance, and citizenship were created, resulting in a new digital world order, fast lifestyles, homogenization of cultures, an information technology boom, and the emergence of service sectors.

India, in the 21st century, also witnessed a gradual transformation of societal values and norms, and rituals and customs embedded in the deep-rooted traditions of the country. While the traditional knowledge remained with the older generations, the newer generations readily adopted a more western and global culture. The lasting features of Indian culture have been collectivism, diversity and tolerance, and the integration of spiritualism and materialism. In globalized India, these aspects disintegrated and were replaced with more westernized values of individualism, egalitarianism, rationalism, capitalism, modern technology, and western scientific thinking. Colonial rule for more than a century enabled a smoother transition into the new cultural era, facilitated by education systems, multicultural exposure and exchange, and industrial growth. The globalized world also accelerated the culture of consumerism that had its roots in the emergence of the 'middle class,' who first promoted the concept of 'luxury.'

During this entire period, the traditional creative industries of India, specifically the crafts sector, underwent a sea change, particularly in terms of their traditional markets and patronage. The traditional local markets dwindled with changes in lifestyles and an increase in access to factory-made products, and the scope of patronage reduced with changing social and economic structures. As the artisans were used to a particular home-based, family-based industry, their setups and traditional ways of working made it challenging for them to respond to the fast-changing world order. Unable to keep up with modern markets and consumers, they were relegated to a marginalized presence in the periphery of the mainstream economy. Their sole identity was that of 'cottage industries,' which implied that they were small, without modern technology, informal, and rural. There was a manyfold increase in the disparity between urban and rural economic development, including capital investments, infrastructure, technology use, social class divisions, education, access to markets and information, ease of doing business, income levels, and standard of living.

Eventually, rural India primarily became a supplier of cheap labor to the modern industries of urban India, in the process losing its rich creative and natural resources and identity. The values of traditional skills and crafts declined in the modern industrial world characterized by fast fashion, use and throw culture, large-scale production, acquisition of more wealth and property, and creation of global consumer brands.

Revival of the traditional handicrafts industry

Parallel to the modern industrial developments, the national government of independent India also recognized the need for the revival and development of the Indian rural cottage industries.

Hence an Industries Conference was organized in 1947. The conference identified several challenges faced by cottage and small-scale industries, including lack of finance, outdated techniques of manufacturing, defective marketing, non-availability of raw materials, and competition from mechanized goods whether imported or locally made. The outcome was a strong recommendation to the Union Government to form a Cottage Industries Board to support the interests of traditional occupations and to elevate them from their crisis (AIACA 2017).

In 1952, the Central Cottage Industries Emporium was established to promote and sell products made by these traditional craftspersons, with the aim of strengthening rural and cottage industries, as laid out in the national policy, for livelihood creation and income generation, equitable resource distribution, and building entrepreneurship. The All India Handicrafts Board (AIHB), also established in 1952 and chaired by Kamaladevi Chattopadhyay, contributed towards a policy framework as well as the institutional and program support that revitalized and nurtured the popularity of handicrafts industries in the country.

The Central and State Governments had recognized at that time that Indian handicrafts are a crucial indigenous economic activity prevalent through the length and breadth of the country, and if supported can contribute hugely to trade and exports, in turn creating wealth and strengthening its national economy. Hence, the AIHB set up national schemes for supporting this sector through promotion, research and design development, technical development, and marketing. During the 1950s and '60s, several institutions were set up across the country, including the Khadi and Village Industries Commission (KVIC), Handlooms and Handicrafts Export Corporation, various emporia in the states, Khadi Bhandar outlets for the sale of hand-spun, hand-woven cloth and handmade products, Regional State Handicraft and Handloom Development Corporations, the Weavers' Service Centres and Design Centres, and the Weavers' Cooperative Apex Societies, all of which constituted a formidable ecosystem in which the traditional handmade creative industries could grow (NCERT 2011). The emporia had also set fair wages and prices for the direct purchase of artisan products and tried to mentor artisans linked to the emporia in marketing and technical aspects. In spite of such efforts, the support of the national government towards the traditional creative sector with subsidies, welfare schemes, and regulatory protections remained largely welfaristic in approach. At the same time, there was an active private sector that realized the potential of Indian handicrafts in the world market, and invested in this industry for marketing, design, capacity building, exports, entrepreneurship. They became strong advocates of this handmade creative industry of India and influenced the government for supportive policies and programs aligned towards the protection and sustainability of the crafts and skilling of the maker communities. With globalization, as the world started coming closer, and with support from national policies to revitalize Indian handicrafts, the producers found new demand and contexts in market spaces outside their villages, and eventually connected with international consumers.

The current scenario

It is clear that there is unmet domestic and international demand for the handcrafted products of India's rural creative industry. In an ideal world, renewed linkages with the world in the internet era should have enabled a resurgence of this traditional creative industry. However, in practice, the industry became fraught with multifarious challenges rooted in the disintegration of village economies and the growth of urban industries, the influence of western culture and modernization, the impact of industrialization, and cultural globalization. Instead of connecting traditional communities to the world, the internet era primarily boosted the urban economic machinery, widening the economic and social divide between rural and urban populations. The capacities of the traditional maker communities remain inadequate for interfacing with the modern markets in a globalized world, in terms of education, business intelligence and capabilities, management, and marketing skills. As these rural creative industries started losing ground, the younger generations stopped imbibing their family traditions and skills as these did not remain profitable, seeking more lucrative employment opportunities elsewhere. Stagnancy in the village industries led to dying skills and capabilities associated with these unique traditional creative occupations.

The non-formal nature of the sector also brought in additional players who are essentially middlemen, who engaged in this industry to take advantage of its structural fragility. Increasing the layers of intermediaries, controlling business, and pushing the artisans further away from fair income and profit, they add to the plight and exploitation of the smaller artisans. Owing to this marginalization, artisans remain delinked from changing design and market trends, information on buyers, markets, and consumer choices, and lack negotiation power for fair business terms.

To address these challenges, various craft enterprises across India have come up with for-profit and nonprofit models to establish innovative and sustainable solutions to the problems faced by the crafts sector. Various development organizations also work to address sustainable livelihood using handicrafts skills, as well as revival and promotion of endangered crafts forms and continuity in practice of the crafts skills by the artisans. In the private sector space, social enterprises that are oriented towards market sustainability are directly working with craft practitioners / organizations / clusters through established and effective marketing mechanisms nationally and internationally, with a focus on socially responsible initiatives. Individual artisans and artisan entrepreneurs themselves are doing business and are striving to scale up and grow. Some of these examples are discussed in Chapters 4 and 5.

The Development Commissioner Handicrafts, Ministry of Textiles, Govt. of India, estimated that India was home to 7 million artisans in 2010–11, of whom about 56% were women. However, unofficial sources estimate that number to be as much as 200 million (IBEF 2021). The village-based, home-based informal nature of the industry, the lack of clear differentiation between traditional practice and professional work, and its sheer scale and diversity have made it difficult to obtain national-level countrywide data on artisans.

More comprehensive region-wise estimates and coordinates of the artisan communities and practitioners are available with various local organizations and enterprises working with artisans and handicrafts. Some of the leading networks and national-level organizations which have enabled and boosted business and growth of this sector over the years include national networks such as the Crafts Council of India, founded in 1964 by Kamaladevi Chattopadhyay, functioning with the support of nine state-level Craft Councils; Dastkar, a society for crafts and crafts people, set up in 1981 to provide capacity building and design support, as well as marketing services through craft bazaars as a platform for rural artisans to sell directly to urban consumers; and Dastkari Haat Samiti, a national association of Indian crafts people established in 1986, working with the objective of providing a common platform to unite craftspersons to work for their own interests, upgrade their skills, innovate their products, and enable them to sell directly to consumers through craft bazaars and urban haats such as the Dilli Haat. There are several notable regional networks as well. The Sasha Association for Craft Producers (SASHA), established in 1978 in West Bengal, is committed to building a fair trade market in the growing domestic market in India. The Self Employed Women's Association (SEWA) is a community-based organization with a membership of over 2 million informal sector women across 14 states and has been working for their socio-economic development over the last four decades. The All India Artisans and Craftworkers Welfare Association (AIACA), established in 2004, was designed as a network of handicrafts and handloom artisans pan India, and works towards sustainable livelihood generation of artisans through enterprise development support and a national market recognized authentic handicrafts certification called Craftmark. These NGOs have institutionalized many good models and best practices, and learning from their work can substantially inform policy. However, the scale of these interventions and their impact is small compared to the size of this sector. The models and contributions of some of these organizations are discussed in Chapter 6.

Among the national institutions are the National Institute of Design, the National Institute of Fashion Technology, and the Indian Institute of Crafts and Design. These institutes build capacities of students who wish to join the crafts-based creative sectors for employment and business. There are also innovative and organized design education initiatives, along with the more conventional and national design institutes. Among them, the more recent institutions are the Somaiya Kala Vidyalaya founded in 2014 in Kutch, and The Handloom School in Maheshwar, which offer sustained, coherent programs for developing artisan entrepreneurs.

There are many handicrafts businesses as well. Fabindia is the largest domestic market retailer in India, sourcing products from over 40,000 artisans. Other notable private crafts businesses include Anokhi and Good Earth which have created niche high-end luxury markets in the handicrafts space. All these crafts businesses have different business models. Among the more contemporary models of artisan-owned enterprises, Rangsutra and Industree Foundation are two leading organizations working towards artisan-owned enterprise development and promotion focused on business development and exports. Moreover, there are newer businesses in the crafts sector including designer-led enterprises and online markets. E-commerce has opened up as a major market space for crafts, with specialized enterprises such as Jaypore.com, itokri, Gaatha.com as well as large e-commerce platforms such as Amazon and Flipkart. Another very important group of private sector stakeholders are designers with exclusive brands (Ritu Kumar, Sabyasachi, Anju Modi, etc.), who have had a significant contribution in linking fashion and changing consumer tastes to traditional crafts, making them relevant and appreciated in contemporary markets. A few of the private enterprises mentioned here are discussed in Chapter 6.

Exports represent a substantial fraction of the rural creative economy. The Export Promotion Council for Handicrafts (EPCH), which is an apex body of handicraft exporters, has over 9000 members who carry out the bulk of handicraft exports from the country. The major marketing channels for exports include wholesalers, importers and distributors, commission agents / sales representatives, department stores, internet-based sales, etc.

However, the scale and diversity of the sector is huge, and such private and public initiatives have remained concentrated in some geographical pockets, leading to unequal growth of this sector.

References

Agrawal, Ashvini. 1989. *Rise and Fall of the Imperial Guptas*. Motilal Banarsidass Publishing House, Delhi.

AIACA. 2017. *"National Handicrafts Policy Report."* AIACA.

Asher, Catherine B., and Cynthia Talbot. 2006. *India Before Europe*. Cambridge University Press, Cambridge, UK.

Boulnois, Luce. 2004. *Silk Road: Monks, Warriors & Merchants on the Silk Road*. Translated by Helen Loveday. Odyssey Books & Guides, Hong Kong.

Chand, Ramesh, Shivendra Kumar Srivastava, and Jaspal Singh. 2017. *"Changing Structure of Rural Economy of India: Implications for Employment and Growth (Discussion Paper)."* NITI Aayog.

Harle, James C. 1994. *The Art and Architecture of the Indian Subcontinent.* Yale University Press, New Haven.

IBEF. 2021. "India's Handicraft Crafts: A Sector Gaining Momentum." www.ibef.org/blogs/india-s-handicraft-crafts-a-sector-gaining-momentum; India Brand Equity Foundation (IBEF).

Kapur, Radhika. 1963. "Small Scale Industries of Mysore." NCAER.

Koga, Masanori. 1968. "Traditional and Modern Industries in India." *The Developing Economies* 6 (3): 300–323.

Maddison, Angus. 2003. *The World Economy: Historical Statistics.* OECD Publishing, Paris.

Mallory, James Patrick, and Douglas Q Adams, eds. 1997. *Encyclopedia of Indo-European Culture.* Fitzroy Dearborn Publishers, London, Chicago.

NCERT. 2011. "Craft Traditions of India: Past, Present and Future." NCERT.

Nilakanta Sastri, Kallidaikurichi, A. 1955. *A History of South India from Prehistoric Times to the Fall of Vijayanagar.* Oxford University Press, Oxford, UK.

Pacey, Arnold. 1993. "Technology in World Civilization: A Thousand-Year History." *Philosophy East and West* 43 (1): 155.

2 The sustainable development approach

In 1987, the landmark Brundtland report titled 'Our Common Future,' published by the World Commission on Environment and Development (WCED), defined *Sustainable Development* as development that meets the 'needs of the present without compromising the ability of future generations to meet their own needs' (WCED 1987; Brundtland 1987). It brought into consideration the need for environmental sustainability along with economic and social sustainability, as they were interlinked and impacted development as a whole. The Brundtland report stated that the severe poverty in the global South, and the unsustainable ways of consumption and production of the global North led to critical environmental threats to the Earth and its inhabitants. Hence, sustainable development emerged as a major concept entwining all development approaches, advocating for long-term solutions towards sustainable resource use and eradication of mass poverty and social inequity, in present and future generations. The understanding emerged that the interlinkages of the three dimensions of sustainable development—economic, social, and environmental—will lead to the desired transformative change.

In the 2019 Global Sustainable Development Report (United Nations 2020), six focus areas have been identified for achieving the necessary changes for a sustainable economy, ecology, and society. These are human well-being and capabilities; sustainable and just economies; food systems and nutrition patterns; energy decarbonization with universal access; urban and peri-urban development; and global environmental commons. The report posits that progress in these identified areas will also lead to higher levels of

DOI: 10.4324/9781003331476-2

well-being and equity for all, and thus, to attain this balance in holistic development, progress needs to be pro-people and pro-nature with localized implementation.

Parallel to the work on achieving sustainable development, the perils of globalization have come to the fore, manifested in a homogenization of cultures, greater inequality, erosion of cultural identities, and environmental degradation. Modernization, which had initially emerged as a concept of linear progression towards a technology-led western lifestyle, has been reinterpreted as a multidimensional process, essentially based on scientific knowledge and local culture, wherein the ultimate purpose of introducing modern scientific knowledge is to achieve a better and more satisfactory life in the broadest sense of the term accepted by the society concerned (Alatas 1972). As discussed in more detail in the preface, culture has been identified as a key driver of sustainable development in terms of generating livelihood and income, developing social and cultural equity, strengthening human and social capital, achieving gender parity, and mobilizing the use of local resources to build modern local economies which are self-sustaining and independent.

Sustainability as a way of life

To understand how traditional creative economies can play a pivotal role in formulating sustainable development models for India, and more broadly for Asia, it is important to note that the philosophy of sustainability has been ingrained in the very existence and culture of its people. This ethos forms an essential basis for leveraging the unique traditional creative economies of the region in the 21st century, not only for economic and social development but also for environmental protection.

For centuries, even millennia, the creative economy of India has been integral to native lifestyles, culture, faith, and societies, which in turn are built on the core components of sustainability. Indian society is traditionally frugal, characterized by recycling and reuse, prudent use of resources, and simple but scientific hand technology for shaping utilities and art. The rural population in India has traditionally lived close to nature and has sustained indigenous knowledge and wisdom of living in harmony with their local biodiversity. This ecological consciousness and ethos of sustainability has continued through the different phases of India's social and economic evolution. Ancient traditions were based on scientific knowledge, efficient resource planning, and collective efficacy, which are demonstrated through centuries-old rainwater harvesting systems unique to the local topography and climate, stepwells for water conservation, consumption of foods that are local and seasonal, adaptability to local vegetation, preventing wastage in all aspects of living, using natural methods of preservation and storage, hand skills and techniques of processing fully local natural materials for building houses, making clothes, cooking, crafting utilities, creating art, application of natural medicines, etc. Embedded in these traditions are the key features of creative industries in India—longevity and continuity, collective identity, tolerance, traditional knowledge and wisdom of nature, and culture passed on through generations. India's traditional handicrafts and handloom sectors have had the attributes of a 'green economy' long before the term existed, with their USP of being local, indigenous, culturally rooted, low energy, and socially and economically enabling for the communities who have held these knowledge systems for centuries.

The principle of sustainability is in fact a cross-cutting value across Asian cultures centered around respect and gratitude for natural resources, and surrounding materiality. The element of humility runs through the local belief systems, and practices of 'reduce, reuse, and recycle' pervade all aspects of life. Some of the interesting policies and practices promoting sustainability and holistic well-being in Asia may set a context for the relevance and significance of creative economies of the global South for achieving sustainable development.

The 'Sufficiency Economy' in Thailand (Piboolsravut 2004) is based on the principles of reasonableness (or wisdom), moderation, and prudence. Two essential underlying conditions are knowledge and morality. The Sufficiency Economy Philosophy (SEP) emphasizes maximizing the interests of all stakeholders and having a greater focus on long-term profitability as opposed to short-term success. The concept was developed by His Majesty King Bhumibol Adulyadej, who emphasized in a 1974 speech that stability of the basic economy should be established prior to investing in the expansion of the industrial sector. The SEP thus focuses on poverty alleviation and community empowerment through the strengthening of the local economy.

The Gross National Happiness in Bhutan is a multidimensional measure of sustained human well-being across nine domains of living standard, health, education, ecological diversity, cultural diversity and resilience, community vitality, time use, psychological well-being, and good governance (Ura et al. 2012). The phrase 'Gross National Happiness' was first coined by the 4th king of Bhutan, King Jigme Singye Wangchuck, in 1972 when he declared that 'Gross National Happiness is more important than Gross Domestic Product.' The GNH principle underlying Bhutan's national economic and social policies takes into consideration a holistic approach towards development that would improve the happiness and well-being of people instead of just focusing on economic growth. This notion of progress has led Bhutan to become the first country in the world to become carbon negative, in spite of tourism being one of its major revenue generators.[1]

Nepal, which is well known for its traditional handicrafts and art forms, has one of the strongest country networks of Fair Trade Organizations recognized nationally and internationally. The Fair Trade Group (FTG) of Nepal is a consortium of grassroots handicrafts organizations and producers that emphasizes transparent and respectful trading partnerships for greater equity in international trade.[2] They negotiate for and promote fair trade as a social movement, enhancing the standards of living of marginalized and exploited producer groups.

The Creative Economy Agency (BEKRAF) in Indonesia is a non-ministerial agency that was established in 2015, by Indonesia's President Joko Widodo to harness the huge potential of Indonesia's creative economy (Harikesa and Afriantari 2019). The Indonesian government was of the view that a specialized agency outside the ministry would be in a better position to enable the growth and strengthening of creative industries through appropriate policies and ecosystem development. The agency covers six functions: research, development and education; access to capital; infrastructure, marketing, facilitation; regulation of intellectual property (IP) rights; inter-governmental relations; and interregional relations. The idea is to boost creative entrepreneurship and attain an economically, socially, and environmentally sustainable economy.

All these countries are similar to India in that they have ancient traditions of indigenous arts and crafts associated with their respective religions, rituals, utilitarian needs, local resources, and cultural practices. The characteristics, ecosystem, and potential are therefore similar across the regions, and this should be viewed as a historical and cultural advantage for pioneering sustainable development in economic, social, and environmental fields.

People-centric inclusive development: Theory and practice

During the second half of the 20th century, the interconnectedness of developmental aspects was demonstrated through research and social experiments by visionary economists and social reformers. Their theoretical discourse and research converge on an approach of people-centric holistic development. In this section, we review some of the major concepts developed and established by them to strengthen the understanding and appreciation of the case studies of creative economy and sustainable development in India that we present in later chapters of this book.

Since the 1930s, the leading idea of progress was based on economic growth as measured by per capita income, GNP, GDP, etc. It was believed that poor countries, that is lower income countries, could overcome underdevelopment by pursuing economic growth. However, many economists such as Amartya Sen and Jean Drèze have pointed out that GDP growth may not always lead to a higher standard of living, particularly in terms of environment, healthcare, and education (Drèze and Sen 2013). Amartya Sen, known for his seminal work in multidimensional poverty and the importance of social and human aspects in meaningful economic development, argued that income growth was just one aspect of development, and that human welfare and human freedom were as important for a healthy economy and society. Hence, the idea of 'welfare economics' pioneered by Sen seeks to evaluate economic policies in terms of their effects on the well-being of the community and the 'richness of human life, rather than the richness of the economy in which human beings live' (Shaikh 2004).

An important concept ideated by Sen was that of 'capability.' He suggested that human capabilities constituted a better indicator for welfare, rather than just commodities or utilities (Sen 1985). He defines capability as the ability of people to function in different capacities in a society such as to enable them to achieve the components or the constitutions of their well-being. According to him, poverty results from the failure to act capably for the achievement of well-being, and therefore a country's wealth lies in the concrete capabilities of its citizens. The development and progress of every country thus depend on multiple dimensions and variables linked to social and human development.

Mahbub ul Haq, a friend and close associate of Amartya Sen, contributed to the path-breaking work of establishing the Human Development Index (HDI), a global statistical framework to quantitatively measure not only economic growth but also human development. His vision of development was also focused on people and their well-being, thus humanizing poverty and underdevelopment. Haq believed that South Asia could become the next economic frontier of Asia if the free flow of rich customs, commerce, and ideas was encouraged, and advocated for a less brutal process of globalization and global institutions that would protect vulnerable people and nations (Haq 2017).

The annual Human Development Reports[3] produced by the United Nations Development Programme (UNDP) have been using HDI since the program was launched by Haq in 1990. The objective was to establish that the people of a nation and their capabilities should be the indicator for assessing a country's development, and not economic growth measured by national income alone. Thus the HDI takes into account 'a long and healthy life, being knowledgeable and having a decent standard of living' (UNDP 2023). However, it does not measure inequalities, poverty, human security, empowerment, etc. In 2010, another measuring tool called the Global Multidimensional Poverty Index (MPI) was developed by the Oxford Poverty & Human Development Initiative (OPHI) and UNDP. The MPI includes the indicators of health, education and standard of living to measure poverty, and is published along with HDI in the Human Development Report.

Although formalized in the second half of the 20th century, these ideas were not completely novel. Rabindranath Tagore, known primarily as a poet and author, was also a visionary social reformer who advocated rural development based on the principles of self-sufficiency. In a talk titled *Swadeshi Samaj* that he delivered in 1904, Tagore criticized the dependency of people on the government for basic amenities, and advocated for a model rural reconstruction that uses emerging technologies to strengthen the lives of people in modern times. He believed that the core reason for poverty and marginalization is lack of education, not only in terms of literacy but an overall advancement of the human consciousness through learning and doing. Tagore's trials of his rural reconstruction model at Silaidaha-Patisar and Sriniketan constituted a pioneering work whose goal was 'the wholesome development of the community life of village people through education, training, healthcare, sanitation, modern and scientific agricultural production, revival of traditional arts and crafts and organizing fairs and festivities in daily life' (Chattopadhyay 2018). He believed that through self-help, self-initiation, and self-reliance, village people will be able to help each other in their cooperative living and become able to prepare the groundwork for building the nation as an independent country in the true sense. He introduced the concept of cooperatives and believed in collective efforts towards local village development. He found immense potential in integrating culture, economic activities, scientific knowledge, and industrial ventures in villages through agriculture and cottage industries, cooperative credit, and self-reliance. Collective good to him was a greater measure of advancement than that of accumulating individual profits.

Mohandas Karamchand Gandhi, an Indian lawyer known for his leadership and advocacy of non-violence during India's freedom movement, was also a social reformer who strongly influenced economic and political thoughts in the country and beyond. He advocated for reconstruction of villages to make them self-sufficient and improve the quality of rural life as part of his philosophy of *Sarvodaya* or 'common good' (Kumarappa 1951). He emphasized a self-sufficient rural economy and polity, with a focus on human dignity and development rather than materialism, as the basis of an equitable, non-exploitative society. Like Tagore, Gandhi promoted a cooperative approach to running village-based industries including farming, crafts, and other cottage industries. To revive and strengthen the village economy, Gandhi affirmed the need for empowering the village population with skills through vocational training and education, and learning by doing and application. According to his vision of a self-sufficient village, the village should produce its basic needs of food, clothing, and some necessary items for daily living, by farming their own land

and producing handcrafted products for their own consumption and use. Beyond that, there would be inter-dependency with other villages and neighborhoods for procuring some products that cannot be produced inside the village. Handicrafts and cottage industries would provide employment to the villagers and facilitate village-based economic independence. Gandhi did not consider industrialization, which exploited rural people who migrated to cities in search of labor work, as a viable path to progress for India. For him, equitable development had to be centered around the villages of India, not its cities and urban centers.

Gandhi also promoted the idea of *Gram Swaraj* or self-governance of villages and believed that the administration and economy need to be decentralized so that strong village institutions can be built, ensuring decent standards of living including cleanliness, public health, comfortable living conditions, education, and increased use of local resources to fulfill village needs. Gandhi's ideas remain relevant in the 21st century, and refocusing on village development is becoming a priority, especially after the COVID-19 pandemic (Ishii 2001; Mishra 2009).

Tagore and Gandhi, two great minds of the 20th century, were friends and admirers of each other. Both talked about the importance of strengthening village institutions and education for the advancement of the human mind.

Muhammad Yunus, an economist who was influenced by Nicholas Georgescu-Roegen and his concept of 'ecological economics' which emphasizes sustainable development, pioneered microcredit with his unique *Grameen Bank* model for empowering village women as entrepreneurs (Yunus 2004). He envisions an economic machinery that does not enable wealth accumulation for only a few but promotes wealth creation for a broad-based, people-centric development of the economy. He advocates for the development of a self-sufficient rural economy that is not dependent on the capitalist economy but is based on local skills and resources, creating material, social, and emotional value for itself resulting in poverty alleviation (Dugger 2006; Bayulgen 2008).

The importance of women's empowerment for societal well-being is evident in the work of each of these scholars, be it through research or social experiments. All of them believed that the process of development needs to be secular, egalitarian, market friendly, and socially uplifting.

Creative cultural industries and inclusive development

The ideas and concepts discussed above converge in their approach towards sustainable development. They consider human and social aspects of development as fundamental to a healthy economy and posit that collective well-being and a people-centric development approach can lead to a sustainable economy as opposed to the idea of advancement as an accumulation of individual profits. Their proponents also argue that human capacities and capabilities in terms of education and skills are more important indicators for assessing a country's growth than its material assets and national income. Thus, these common tenets put forward a development theory that merges economic, social, and human welfare and empowerment to assess sustainable development of any economy and society.

The unprecedented COVID-19 pandemic shook the existing paradigm of economic growth and made the world realize the importance of local, self-sufficient, self-reliant, and human creativity-based occupations. Not only was it evident that rural economies were more resilient and recovered from the economic shock of lockdowns faster, but the plight of migrant workers stuck in Indian cities under untenable living conditions highlighted the fragility of the urban economy whose apparent growth is dependent on the exploitation of cheap labor from the rural hinterland, especially as regulatory policies lag behind in the fast-evolving world of the gig economy. Dialogues re-opened on alternative economic models that would be based on well-being, resilience, collective good, and strong cultural identity. Development of local economies and strengthening local development agendas have been prioritized across multiple sectors such as business and entrepreneurship, industry and employment, governance, environmental protection, social welfare, and infrastructure development.

India's immense cultural and geographical diversity, primarily rural population covering thousands of localized creative communities, rich history of ancient civilizations and cultures, experiences of wide cultural diffusions, and shared heritage with bordering countries make it a unique place for understanding practical aspects of the creative economy in achieving sustainable development.

We conclude this chapter with a summary of how the traditional creative economy resources in India can help achieve people-centric development goals and the related sustainable development goals (see preface) of 'no poverty' (SDG 1), 'good health and well-being' (SDG 3), 'quality education' (SDG 4), 'gender equality' (SDG 5), 'clean water and sanitation' (SDG 6), 'decent work and economic growth' (SDG 8), 'industry, innovation and infrastructure' (SDG 9), and 'peace, justice, and strong institutions (SDG 16).'

The developmental goals of human welfare, human freedom, and well-being of communities are linked to SDGs 1, 3, 5, 8, and 16. The resources and conditions present in Indian society to achieve those are:

- the age-old traditional creative skills and traditional creative occupations that engage millions of people;
- village-based industries having a familiar and harmonious home-based work setup and family support, providing social and emotional wellness;
- the integral engagement of women in traditional creative economies in which they fulfill pivotal role in production, processing, and generation of creative products;
- traditional collectives of homogeneous creative communities which form strong self-reliant societies and are inclusive of all community members in specialized roles of creative production.

The goal of concrete capabilities of communities and citizens, and overall advancement of the human consciousness through learning and doing to achieve a healthy sustainable economy and society, are linked to SDGs 1, 4, 5, 8, and 9. The resources and conditions present in India to achieve those are:

- traditional creative occupations and a rich history of trade and commerce in cultural goods;
- traditional creative skills honed and transmitted through generations resulting in skilled creative producers;
- indigenous knowledge of local ecosystem, local resources, natural environment, and art and craft skills for daily sustenance (construction and architectural skills, weaving skills, cooking skills, crafting utilities, cultivation, bio-cultural landscape management), inherent in the communities' way of living.

The goal of humanizing poverty and underdevelopment, by integrating advancement in health, education, and standard of living are linked to SDGs 3, 4, 6, and 8. The resources and conditions present in India to achieve those are:

* an age-old culture of creative production connected to the emotional well-being, dignity of work, and pride in one's culture, apart from economic benefits;
* emotional well-being leading to personal and social well-being, aspirations for improved standard of living, and need for education to become equal participants in the globalized modern world;
* respect and recognition for specialized and unique local creative skills based on local resources that create opportunities for creative entrepreneurship contributing to the overall advancement of practitioner communities, overcoming marginalization and social divides.

The goal of self-sufficiency and rural reconstruction for the wholesome development of community life is linked to SDGs 1, 3, 5, 8, 9, and 16. The resources and conditions present in Indian society to achieve those are:

* cohesive communities of Indian village-based traditional creative occupations, bearing the same skills, knowledge and cultural contexts, are ready units for building and enhancing local economies, and wholesome community development. In rural India, the specialized practitioner communities producing specific creative products constitute the nucleus of self-sustaining independent economies based on local skills and resources.

The commonalities in the ideas put forward in this chapter reiterate the need for an alignment of current practices along the principles of people-centric economic development theories. It also provides a foundation for the remainder of this book, which delves into specific case studies on the development and strengthening of traditional creative industries that resonate with post-COVID development approaches. These concrete examples demonstrate the concepts of 'capability,' 'human capital as wealth of a nation,' 'enrichment of human life,' 'social justice,' 'inclusiveness,'

'international cooperation,' and 'people-centered development.'
Although the case studies are not directly motivated by these theories, it is useful to view them through the lens of these ideas, especially if the eventual goal is to replicate or adapt and scale up these models.

Notes

1 https://ophi.org.uk/policy/gross-national-happiness-index/ (accessed May 2023).
2 www.fairtradegroupnepal.org/ (accessed May 2023).
3 https://hdr.undp.org/about/human-development (accessed May 2023).

References

Alatas, Syed Hussein. 1972. *Modernization and Social Change: Studies in Modernization, Religion, Social Change and Development in Southeast Asia*. Angus; Robertson, Sydney.
Bayulgen, Oksan. 2008. "Muhammad Yunus, Grameen Bank and the Nobel Peace Prize: What Political Science Can Contribute to and Learn from the Study of Microcredit." *International Studies Review* 10 (3): 525–47.
Brundtland, Gro Harlem. 1987. "Our Common Future—Call for Action." *Environmental Conservation* 14 (4): 291–94.
Chattopadhyay, Madhumita. 2018. "Rabindranath Tagore's Model of Rural Reconstruction: A Review." *International Journal of Research and Analytical Reviews* 5: 142–46.
Drèze, Jean, and Amartya Sen. 2013. *An Uncertain Glory: India and Its Contradictions*. Princeton University Press, Princeton and Oxford.
Dugger, Celia W. 2006. "Peace Prize to Pioneer of Loans to Poor No Bank Would Touch." www.nytimes.com/2006/10/14/world/asia/14nobel.html.
Haq, Khadija. 2017. "Mahbub Ul Haq: Pioneering a Development Philosophy for People." https://blog.oup.com/2017/09/mahbub-ul-haq-philosophy-economics/.
Harikesa, I Wayan Aditya, and Rini Afriantari. 2019. "Creative Industries 4.0: BEKRAF Economic Program and Boundary Concept." In *The 1 International Conference on Innovation of Small Medium-Sized Enterprise (ICIS) 2019*. Universitas Pasundan Bandung, Indonesia, 209–9.
Ishii, Kazuya. 2001. "The Socioeconomic Thoughts of Mahatma Gandhi: As an Origin of Alternative Development." *Review of Social Economy* 59 (3): 297–312. www.jstor.org/stable/29770117.

Kumarappa, Joseph Cornelius. 1951. *Gandhian Economic Thought*. AB Sarva Seva Sangh Prakashan, Bombay.

Mishra, Harindra Kishor. 2009. "Relevance of Economic Ideas of Gandhi in 21st Century." In *Economics of Mahatma Gandhi: Challenges and Development*, edited by Anil Kumar Thakur and Mithilesh Kumar Sinha. Indian Economic Association. Deep & Deep Publications, New Delhi, 508 pages.

Piboolsravut, Priyanut. 2004. "Sufficiency Economy." *ASEAN Economic Bulletin* 21 (1): 127–34.

Sen, Amartya. 1985. *Commodities and Capabilities*. North-Holland, Amsterdam.

Shaikh, Nermeen. 2004. "Amartya Sen: A More Human Theory of Development." https://asiasociety.org/amartya-sen-more-human-theory-development.

UNDP. 2023. "Human Development Index (HDI)." https://hdr.undp.org/data-center/human-development-index.

United Nations. 2020. "Global Sustainable Development Report 2019: The Future Is Now – Science for Achieving Sustainable Development." UN. https://sustainabledevelopment.un.org/content/documents/24797G SDR_report_2019.pdf.

Ura, Karma, Sabina Alkire, Tshoki Zangmo, and Karma Wangdi. 2012. *A Short Guide to Gross National Happiness Index*. The Centre for Bhutan Studies, Thimpu, Bhutan.

WCED. 1987. *Our Common Future*. Oxford University Press, Oxford, New York.

Yunus, Muhammad. 2004. "Grameen Bank, Microcredit and Millennium Development Goals." *Economic and Political Weekly* 39: 4077–80.

3 The regional context

An overview

As discussed in the previous chapters, India's handicraft industry, taken together, is one of the largest and most diverse in the world and bears the potential to become a multi-billion-dollar creative industry. The business ecosystem centered around this industry, particularly handicraft design, production, innovation, and marketing, has been quite strong among urban designers, contemporary fashion and design brands, and exports. However, the key to making this a sustainable industry is self-sufficiency, by enabling a business environment and providing resources for the actual rural artisans and their home-based traditional work. It was realized by socially responsible brands and entrepreneurs long back that the inclusion of artisans in the business process and empowering them to produce market-oriented products was critical in order for the actual legacy of Indian handicrafts to be sustained and brought to the world market successfully. These leading brands and entrepreneurs have made significant contributions towards the development and growth of the handicrafts market in India and abroad.

Exports have traditionally represented a significant market for Indian handicrafts. According to data reported by EPCH,[1] handicrafts exports (excluding carpets) grew from INR 713 crore (USD 150 million) in 1990–91 to INR 33253 crore (USD 4 billion) in 2021–22, with the USA accounting for 42% of this amount. A relatively recent but significant development in the sector is the proliferation of e-commerce markets in the handicrafts business. There are many online marketplaces and stores dealing with handicrafts; while some specialize in only handicrafts, or specific

DOI: 10.4324/9781003331476-3

segments of handicrafts such as textiles or jewelry, others like Amazon and Flipkart are large online markets selling handicrafts as one of many product segments.

E-commerce had gained momentum since the mid-2000s, but post pandemic it has brought a major shift in consumer behavior towards online purchases. There is an argument that in the handicrafts sector, touch and feel is important as the diversity of materials and how they are processed and productized are important factors for consumer choices, and hence physical outreach is essential. In reality, it is found that once an enterprise has built trust in the market towards its brand quality and integrity, consumers start purchasing online. The e-commerce platforms offer direct artisanal products, exclusive curated collections which are co-created with the artisans, and also host well-known artisan brands on their platforms.

Handicrafts have become popular as souvenirs and gifts in the travel and tourism industry. Although this market always existed in destinations offering heritage tourism, the trend has spread to other types of tourism as well. Experiential tourism has expanded to include not only natural and historical places, but also traditional art and craft villages where one can experience unique handmade processes of the crafts or textiles, know about the community's lifestyles, and buy products directly from the makers' own habitats. Today, tourism includes not only the handicraft products as souvenirs but also the cultural experience associated with those products as a direct offering by the traditional craft communities. Largely linked to the tourism industry is the specialization of interior design which is engaging traditional artisans to creatively enhance public and business spaces, hotels, and cafes, etc.

Corporate gifting is another market segment which has started investing in artisanal handicrafts for strengthening their brand image and identity, and adding uniqueness to their branding.

As different types of markets have emerged, the artisans have also grown and evolved to become more creative, innovative, and modern, stretching their imagination from traditional handicraft products to diverse handicraft applications in home decor, interiors, architecture, etc. International designers and brands have increasingly come forward to collaborate with Indian artisans to produce high-value handmade products fitting global aesthetics and procure unique exclusive handcrafted items from the country. Today, Indian handicrafts have a very high visibility and awareness around the world, yet it struggles to become a full-fledged industry that is inclusive, people-centric, and self-sufficient in its own right. The case studies presented in the following chapters represent models that have been successful in enabling more equitable and sustainable business practices based on traditional craft industries. The case studies are from three specific states in India, each characterized by unique circumstances that have led to a remarkable diversity in handicrafts, and traditional creative industries. The remainder of this chapter will briefly describe these three states, West Bengal, Gujarat, and Rajasthan, focusing specifically on their history, topography, and natural resources, to set the context for their handicrafts industries.

West Bengal

Geography

West Bengal is a state in eastern India endowed with diverse natural and cultural resources. The state is characterized by high-altitude landscapes of the eastern Himalayas, foothills and valleys, riverine regions, dense forests, arid regions, and coastal areas. It shares its borders with Bangladesh in the east, and Nepal and Bhutan in the north. It also borders the Indian states of Odisha, Jharkhand, Bihar, Sikkim, and Assam.

The Himalayan hill region and foothills in the north are covered with temperate, sub-tropical, and tropical forests, the rugged *Rarh* region in the west is a land of red soil, and the Ganges delta and the coastal Sundarbans border the Bay of Bengal in the south. Hence, the people of the state have distinct lifestyles leading to many different traditions and handicrafts. Its main economy is based on agriculture, but it also has many different micro and medium industries, of which cottage industries constitute a significant portion. The local vegetation, soil types, and weather conditions dictate the use of local natural materials, types of utilities crafted, and the range of cottage industries associated with people's traditional practices.

History

In the past, the multitude of rivers and the coast of the Bay of Bengal in the state offered the opportunity of easy communication with its hinterlands for internal trade. In the ancient period, West Bengal was part of a much larger territory including present-day Bangladesh and parts of Bihar and Odisha. The undivided Bengal region was part of several ancient pan-Indian empires, including the Vangas, Mauryas, and the Guptas. Bengal had established trade relations with Southeast Asia since the beginning of the Christian era, or even earlier. Endowed with large rivers, the world's largest delta, and a narrow land bridge connecting the subcontinent with Southeast Asia, Bengal traded with mainland Southeast Asia, Burma, Sri Lanka, and others, both by overland and maritime routes. South Bengal held the most important position in sea-borne trade which is believed to have continued from the early Christian era till the 11th–12th century CE. These trade networks were more than just avenues of product transactions and fostered relationships of merchant communities and cultural exchanges. During the Bengal Sultanate, Bengal was a major trading territory in the world, often referred to by the Europeans as the 'richest country to trade with.' Mentionable among the contemporary products of this region was Muslin, a fine cotton cloth exclusive to the state, which used to be produced by the local weavers of South Bengal and was one of its major export items at one point of time.

Bengal also experienced cultural diffusion with the Mughals, as the state was brought under the Mughal Empire in 1576, and nominally remained a part of the Mughal regime until the British East India Company effectively seized control in 1757. From 1772 to 1911, Calcutta was the capital of British India. In the later part of this period, Calcutta became a center of the Indian independence movement, as well as one of the richest centers in the country in terms of artistic and intellectual pursuits. The fallouts of the freedom movement affected the native people of the state, as they experienced partition, conquests, religious turmoil, migration, tribal agitations, cross-border movements, westernization, etc., all of which in turn shaped its art and craft traditions over time.

Handicraft traditions

West Bengal has the wealth of a wide array of traditional handicrafts and folk art forms integrally linked to distinctive cultures of the people. Many different types of crafts have thrived in the state including cotton and silk textiles, jute, metal, pottery, unique local natural fibers, folk paintings, and wood craft. More than 60% of the population is rural, and about 5.8% of the total population is tribal as per the 2011 Census.[2] Tribal communities, which have their own indigenous craft traditions, are distributed over all districts, although a few districts have larger concentrations. In most cases, an entire village of a region specializes in a particular craft tradition with styles and techniques that have lived through generations. As per the 4th All India Handloom Census (2019–20), West Bengal has 3,66,656 handloom weavers and 2,64,791 ancillary workers of the handloom industry. As of 2021, it had 2,70,518 artisans enrolled in the PAHCHAN artisan ID card initiative, second only to Uttar Pradesh.[3]

Recognizing the fundamental role of entrepreneurship in developing and strengthening a traditional skill-based creative economy, the Government of West Bengal started investing in developing Rural Craft and Cultural Hubs in partnership with UNESCO in 2013. Under this partnership, the Department of Micro Small & Medium Enterprises and Textiles (MSME&T), Government of West Bengal, has started promoting the craft heritage of West Bengal to bring about sustainable livelihoods for the maker communities. Community-led and managed Rural Craft Hubs have been developed through skill development, product innovation, entrepreneurial skill building, infrastructure development and technology upgradation, wider market outreach, cultural tourism promotion, etc. The objective has been to establish a vibrant creative economic sector, thereby ensuring socio-economic inclusion of the rural poor.

Promoting rural creative entrepreneurship has been taken up as a priority, for meaningful employment of youth in particular, through micro, small and medium enterprises.[4] Handicrafts of Bengal have been identified as one of the high-potential, dynamic, dominant, and vibrant segments of West Bengal's economy. The sector represents an economic lifeline for many vulnerable sections of the society, with women accounting for around 50% of the artisans, and a significant number belonging to the marginalized or minority sections of the society. It provides low-cost, green livelihood opportunities to more than 550,000 men and women. Every district of the state has its own characteristics and features with offerings of unique handicrafts and communities of practitioners that hold the promise to emerge as a creative industry.

Gujarat

Geography

The state of Gujarat is located in western India, on the coast of the Arabian Sea, bordered by Pakistan and Rajasthan in the north east, Madhya Pradesh in the east, and Maharashtra and the Union territories of Diu, Daman, Dadra, and Nagar Haveli in the south. Its coastline, formed by the Arabian Sea on its west and south west borders, is the longest in the country at about 1600 km, most of which is part of the Kathiawar peninsula. The state has mountainous terrains covered with dry and moist deciduous forests, grasslands, salt marshes, wetlands, rivers, and marine ecosystems of coral reefs, estuaries, gulfs, and mangroves. Of the different mountain ranges, the Girnar hills are the most prominent. The state has several beaches, dense forests, as well as arid rugged terrains and is home to a large tribal population, about 14.8% of the state's population.

Gujarat is known for being one of the most prosperous states with significant economic sectors of agriculture, dairy, industrial production, and exports. Owing to its geographic location and mercantile history, the state contributes a major share to merchandise exports. Gujarat's local cultural practices and handicrafts enterprises are shaped by the diversity of people and local natural resources, challenging terrains, and a rich history of multi-cultural diffusion, trade, and commerce for centuries.

History

There are 23 sites of the ancient Indus Valley civilization in Gujarat, more than any other state in India. Among these,

Lothal is believed to be one of the first seaports in the world. The coastal cities of Bharuch and Khambhat were ports and trading centers during the Maurya and Gupta rule. Gujarat has a history of conquests, international trade in the pre-British and British regimes, and represents a confluence of different cultures and communities. Various Indian empires, such as the Mauryas, Western Satraps, Satavahana dynasty, Gupta Empire, Chalukya dynasty, Rashtrakuta Empire, Pala Empire, Gurjara-Pratihara Empire, Maitrakas and the Chaulukyas, have had significant cultural, social and economic influences on its people.

Historical evidence suggests that Gujarat was engaged in trade and commerce with Egypt, Bahrain, and Sumer in the Persian Gulf from 1000–750 BCE. This trade continued into the medieval period, as evidenced by fragments of printed cotton, one of the crafts that Gujarat is famous for even today, found in Egypt from various eras spanning the 10th to 16th centuries. The merchants of Gujarat had been specializing in overseas trade and commerce for many centuries, thus making the state well known for its wealth and prosperity. Gujarati merchants are believed to have earned an international reputation during that period, for their commercial acumen and enterprising nature, which attracted merchants from overseas as well. Gujarat's long-term experience in trade also led to the formation of a diaspora community of merchants from Gujarat, outside the state, and across the world that further added to their long-distance trade potential.

From the 11th century onwards, Gujarat came under Islamic influence, when Arab merchants settled along its western coast. In the 14th century, it was annexed by the Delhi Sultanate but was ruled by the independent Gujarat Sultanate from 1407, until it was conquered and incorporated into the Mughal Empire by Akbar in the 1570s. The Surat port, which was the only Indian port on the western coast of the country, became the main trading port of India during the Mughal rule. By the 1600s, the Dutch, French, English, and Portuguese had all established bases along the coast of this region. The Maratha Empire dominated most of Gujarat in the 18th century, before the British East India Company took control in the early 19th century.

Post independence, Gujarat emerged as an industrial hub with an inherited entrepreneurial spirit that made it resilient, in spite of being hit by major natural disasters and political turmoils.

Handicraft traditions

Gujarat has a rich and vibrant tradition of diverse handicrafts including needle work, tie-dye, block printing, weaving, jewelry, pottery, metal-craft, and woodwork. Business attitude and acumen being integral to the state's culture, artisan communities have grabbed opportunities to establish businesses in traditional handicrafts, which are combined with modern ethos and design sensibilities.

Kachchh, located on the coast of the Arabian Sea, is one of the largest and culturally richest districts of Gujarat. It is home to a wealth of biodiversity and a large population of nomadic communities bearing the tradition of exquisite craftsmanship. The crafts of Kachchh were traditionally made by the rural families for their own ritualistic or daily use, as well as for barter trade and cultural transactions from age-old times. For instance, the famous mirrored embroideries of Kachchh were used for marriage transactions and for fulfilling other associated social obligations in the community.

The uniqueness of Kachchh not only lies in its large diversity of local crafts but the fact that such diversity is naturally concentrated in one region. Kachchh has many traditional crafts-based artisan clusters providing livelihood to the maker communities. Every village has a different and unique style of crafting these products, portraying distinctive identities. Embroidery is one of the most popular crafts in Kachchh, with more than 40 styles of embroideries being practiced in the region, representing Kachchh's diverse cultures, communities, and landscapes. Community identity and pride are deeply embedded in the region's craft culture. Until three decades ago, each craft's production process, from the sourcing of raw materials to the sale of finished products, was completely local and practiced within the artisan communities.

In 2001, a massive earthquake hit Kachchh, resulting in significant destruction of the place and loss of life. Although the people's regular livelihoods including crafts were severely impacted, the response to the devastation also led to the emergence of new opportunities, initiatives, and potential. The history of the state has shown that its people have an inherent strength and spirit of resilience and energy for reviving business and livelihoods. In Kachchh alone, handicrafts provide primary income to about 60,000 artisans. Post earthquake, rebuilding of the region was supported by the local wealthy and enterprising people. For crafts, a collaborative platform called 'Kutch Craft Collective' was formed by five leading local craft organizations—Shrujan, VRDI, Kalaraksha, Qasab and Khamir—who undertook a holistic development approach to revitalize the local craft traditions and establish a unique identity for authentic crafts of Kachchh. Eventually, it became one of the most prominent handicraft business hubs of the country, with substantial international exposure. Today, the handicrafts industry of Kachchh is the most successful in India in terms of artisan entrepreneurship, innovation, revitalization and contemporization of traditional crafts, involvement of youth from artisan families in their traditional craft businesses, empowerment of women and marginalized artisan communities, global outreach, and international business of the rural artisans.

Rajasthan

Geography

Rajasthan is a state in northwestern India bounded to the north and northeast by the states of Punjab and Haryana, to the east and southeast by the states of Uttar Pradesh and Madhya Pradesh, to the southwest by the state of Gujarat, and to the west and northwest by the provinces of Sindh and Punjab in Pakistan. The Aravalli mountain range cuts across the state. In the far west and northwestern part lies the Thar desert. The dry arid land of the west shifts to comparatively fertile land toward the east. The Aravallis form Rajasthan's most important watershed. The climate of the state is extreme with summer temperatures rising to 50 degrees Celsius. While hot winds and dust storms occur in the summer, especially in the desert region, the eastern parts have more humid weather. In the winter months of December– January, temperatures drop to 4 degrees Celsius. The main vegetation of western Rajasthan is scrub jungle. Trees are scarce and are found mostly in the eastern part of the state in the form of small, scattered forest areas in the Aravallis. The two important rivers are Chambal and Luni, the former draining the northeast part, and the latter flowing in the west. Although the state is scarce in water, agriculture has been one of the main occupations in Rajasthan's economy. It is also a major wool-producing state owing to its large population of livestock. Rajasthan is very rich in minerals, which include emeralds, garnets, and silver ore, contributing to its well-known gems and jewelry industry. Other than the manufacturing of various industrial products, Rajasthan is renowned for its diverse handicrafts and textiles. Tourism is one of the most significant industries of the state, contributing to a major share of the state's economy. According to the official portal of Govt. of Rajasthan, about 75% of the state population is rural with about 18% from scheduled castes and about 14% from scheduled tribes. Among the native indigenous communities are people who are traditionally artisans and travelling traders, as well as those who practice farming, cattle-breeding, archery, metal work, etc.

History

Rajasthan has a rich history which is believed to be about 5000 years old, having traces of human settlement from the Indus Valley Civilization. Before the Rajput clan came to power in Rajasthan during the 9th century CE, the region was under the rule of several empires. The dominant ones were those of the Mauryas, Malavas, Arjunyas, Yaudhyas, Kushans, Saka Satraps, Guptas, and Huns. Many Rajput dynasties came into power between the seventh and eleventh centuries. From the 13th century, parts of Rajasthan came under the Muslim rule although the Rajputs retained a degree of independence over the next four centuries. Eventually they were defeated in the 16th century by Akbar, who established diplomatic relationships with the Rajput rulers, leading to an intermingling of cultures. In the early 18th century, some of the Rajput states continued to flourish under the Maratha Empire. In 1857, the British crown formally colonized and started their rule in India, turning most Rajput states into their allies, allowing them to continue their independent existence subject to certain political and economic constraints. Under British rule, the separate Rajput states signed a treaty and came under an umbrella called Rajasthan.

The erstwhile states of Bikaner, Jodhpur, and Jaisalmer, which are located in the barren arid desert region of western Rajasthan, could not earn much from land revenue as agriculture was unremunerative due to the climatic and geographic conditions. However, several trade-routes passed through western Rajasthan connecting the then princely states with adjoining states as well as with Sindh and Ahmedabad. Caravans with convoys of merchandise would pass through these routes and often had to halt at safe places for food and shelter. This was an opportunity which the local communities of western Rajasthan utilized to levy taxes and transit duties earning the state income. It has been documented that about 400 years ago, Jaisalmer in Rajasthan became one such stop-over place for merchant travellers. The Thar desert is also believed to have been part of the ancient Silk Route network.

The communities in Rajasthan also had strong business acumen and were enterprising traders. They did not limit themselves to their homeland, but eventually moved to Calcutta, Bombay, and other parts of the country, initially for new business opportunities in trading, and subsequently in manufacturing.

Handicraft traditions

Rajasthan boasts a plethora of traditional crafts, jewelry, and textiles of a large variety. Often termed the treasure trove of handicrafts in India, Rajasthan is home to numerous traditional handicrafts that have survived and evolved through generations. These include tie and dye work, block printing, embroidery, enamelled gold and silver jewelry, leather crafts, blue pottery and clay pottery, marble work, carpets and rugs, ivory craft, miniature paintings, metal and wood carvings, and embossed brass. The combination of a royal heritage along with strong business communities have contributed to the region in many ways, with heritage monuments and palaces, exquisite artwork, and traditional handicrafts, together constituting a vast cultural asset base for creative and tourism industries. These handicrafts and cultural heritage assets have also developed through time with multiple influences and patronages, bringing out the best of skills and creativity in the makers.

Conclusion

The case studies presented in the following chapters describe social experiments in these three states, by different entities, who have worked towards transforming existing and traditional creative skills and cultural practices into resources for holistic village or area development. They illustrate the emergence and development of different rural creative industries across different geographies and cultures. These examples are by no means exhaustive, and have been identified, based on first-hand experience, as representative examples to highlight a few successful models that also showcase the principles of people-centric economic development discussed in Chapter 3. Some of these initiatives have been undertaken through a public-private partnership model by the state government. Others are NGO-led projects carried out in partnership with individual designers, craft experts, and social entrepreneurs, as well as private craft business enterprises in urban India.

Notes

1 https://epch.in/policies/exportsofhandicrafts.htm (accessed May 2023).
2 https://wb.gov.in/about-west-bengal-facts-figures.aspx, https://adibas ikalyan.gov.in/html/st.php (accessed May 2023).
3 https://pib.gov.in/PressReleasePage.aspx?PRID=1742840 (accessed May 2023).
4 https://ruralcrafthub.com/crafts/what-are-the-crafts/ (accessed May 2023).

4 Self-sufficient village economy

As the majority of the Indian population is rural, it has been reaffirmed time and again by theorists and practitioners alike that upliftment of the marginalized rural sections of the society and development of strong village institutions targeting economic independence are essential to impact the economic and social development of the country as a whole. The case studies in this chapter focus on the economic self-sufficiency of villages in rural India in terms of capacity to generate resources for subsistence and improvement of individual and community life.

The primary requirements of a self-sufficient village community are local income generation based on local resources, distribution of income across multiple occupations of the village communities, collective resilience, and well-being of the village people rooted in their traditional culture, practices, and relationships. In the modern globalized world, such an idealized self-sufficient village economy, disconnected from the rest of the world, is impractical. The case studies presented are not fully self-sufficient village economies in this orthodox sense, but they illustrate strong village-based community-led economies that have made the villages economically self-supporting and competent to independently operate in the larger economic system.

Rural craft hubs in West Bengal

The development of rural craft hubs in West Bengal is part of a project undertaken by the state's Department of MSME&T along

DOI: 10.4324/9781003331476-4

with UNESCO, New Delhi, with an NGO, Contact Base, as their implementation partner.[1]

The vision of the project is to transform different villages of the state, which have been traditional abodes of indigenous handicraft communities, into active, vibrant cultural hubs based on the local handicraft-based creative industries. The aim has been to equip the communities with capacities for business and entrepreneurship in their traditional handicrafts to cater to the modern markets. These capacities entail revitalization of the crafts, most of which have been dying due to lack of market viability, by imbibing modern design sensibilities through exposure and training, professionalizing the creative practices, and establishing enterprises suitable to carry out business. One of the fundamental principles of this process is to revive, safeguard, and maintain the uniqueness of the traditional skills and practices in terms of choice of local raw materials, continuation of indigenous knowledge, adherence to traditional folk aesthetics, and conserving the handmade process.

The model applied by Contact Base involves mobilization of the rural creative communities to become key stakeholders in the process of their development, exposure to the outside world of art, and innovation through exchange collaboration projects, capacity building, and participatory planning with the rural craftspersons about revitalization of the art form, organizing the village craftspersons into a collective community enterprise, and facilitation of direct market linkages for these creative communities and enterprises (Bhattacharya and Dutta 2022). A resource center has been established inside each of the artisan and artist villages that are part of the project, on community land, equipped with IT and internet, and housing a community museum. The enhancement of the community skills and capacities and strengthening local resources and infrastructure with modern facilities have enabled the growth of these villages into successful cultural hubs and destinations for creativity, and creative businesses.

A few of their success stories are as follows.

Patachitra folk art of Naya village: The Patuas or Chitrakars are a community of folk artists of Bengal who are painters, lyricists and singers, and the traditional bearers of a unique visual-oral art form. They paint on scrolls which are called *Patachitra* (Bajpai 2015). These scroll paintings traditionally illustrate a story through the pictures, accompanied by narration in the form of songs. These songs are called *Poter Gaan*. This tradition dates back at least to the 13th century. Traditionally, Patuas made their living from alms, in return for their door-to-door performances which served as entertainment for well-off and middle-class villagers. However, over the years, this traditional practice declined due to a lack of patrons, and the onset of other entertainment channels such as television. With loss of earnings, these artists were forced to give up painting and take up menial jobs to survive. Many migrated to cities for labor work. The Patachitra village of Naya in the West Midnapore district of West Bengal, although traditionally a village of Patuas, was virtually unknown and marginalized about a decade ago. Although renowned scholars, art historians, artists, and designers had been working with specific Patachitra artists from this village, those were specialist-to-artist engagements involving artist mentorship, with the goal of individual growth towards excellence and renown.

In 2005, Contact Base started an intervention to revive Patachitra and turn it into a sustainable and dignified source of income for these artists. As most of the artists had stopped painting long back, mobilizing them to return to their traditional art, which they considered a dead end, required intensive and long-term engagement with the community to build trust and community ownership. A field assessment carried out at the beginning of the intervention showed that although some of the men still continued to fulfil occasional orders and attended a few local fairs, the women did not, in spite of being highly skilled artists themselves. There were hardly 20 practicing artists in the village who earned some income from their Patachitra paintings.

As painting Patachitra is a family tradition, and a great way of spending free time, the women were found to paint in between household chores as a pastime. The children learned the art form organically in the process. Once the community of painters became stakeholders and partners in the project interventions, new capacities, new markets, meaningful promotion, art appreciation, consumer education, revival of painting with traditional natural colors and old themes and subjects of Patachitra art led to their economic, social, and cultural empowerment. Village development in terms of ensuring children's formal education, setting up of necessary infrastructure, sanitation, and workspace improvements were critical interventions spearheaded by the community.

A resource center inside the village housing a community museum enabled the village to grow as a dynamic center of Bengal Patachitra. The resource center started being used as a space for training, exchange and collaboration, skill upgradation, storage, and display. The youth learned to use digital media and manage the village community museum showcasing the history and evolution of their art form.

To promote the village as a whole as a Patachitra hub, and not just individual artists, Contact Base also initiated an annual village festival called "Pot Maya." To strengthen the importance of the traditional habitat of the Patachitra artist community, and the authenticity of Bengal Patachitra, the tag of Geographical Indication (GI) for Naya was also applied for. GI status is a mark for distinguishing products from a particular geographical origin having qualities that are due to that origin. GI ensures and protects the collective rights of the producer community on their products and process of production. In 2018, Bengal Patachitra was accredited with a GI tag.

Today the Patuas of Naya constitute a distinguished artist community who earn a decent income from their art form and enjoy a sustainable livelihood. They work from their homes and often travel the world to train, collaborate, work, perform, and exhibit. There is a global recognition of the artists and their art form which has brought pride and dignity within the community, inspiring them to choose Patachitra as their primary source of income. The younger generations are also choosing Patachitra as a career option and are strengthening their household businesses further.

These rural artists have overcome marginalization and are respected for their creative skills. From about 20 practicing artists in 2005, there are now more than 250 practicing rural artists from about 70 households in the village. There are many women artists who are world-renowned and are taking forward their creative enterprises. Empowerment and leadership of the women artists have in turn contributed to the overall well-being of this community and the general upliftment of their village. It has improved the education of children, enhanced facilities for health care and sanitation, and has led to socio-economic empowerment of the village women.

Since 2010, the artists have collectively taken the lead to continue the annual three-day village festival, Pot Maya, to celebrate their village as an active Patachitra hub, attracting local and global audiences. Geographical Indication has further helped to put the village on the cultural destination maps of Bengal, creating a special identity of the place and the people. The Naya village has grown into a unique cultural tourism destination and has been documented as 'best practice' by the UNWTO.

Dokra metal craft of Bikna village: Dokra is an age-old metal craft of eastern India (Chakroborty, Chatterjee, and Choudhury 2021). Bikna in the Bankura district of Bengal is a hamlet of about 60 Dokra artisan families who settled here in the 1960s. Dokra is a unique craft of lost wax metal casting, used to create a wide repertoire of figurines, deities, decorative pieces, jewelry, pots, etc. It is believed that Dokra is a primitive technique which dates back to the Indus Valley civilization. The Dokra artisans of Bengal belong to the Karmakar community, who were traditionally blacksmiths.

Bikna village was found to be in a poor condition in 2013. Due to a lack of opportunities and exposure, the Dokra business was controlled by only a few artisan traders who commissioned work to others on a wage basis, with the average income of such artisans barely meeting their minimum subsistence needs. They often borrowed from local moneylenders to supplement their income, and in turn got exploited to become bonded laborers. Apart from poverty and illiteracy, the community suffered from poor health. The old furnaces used for Dokra production emitted thick smoke resulting in hazardous air quality.

Contact Base applied the same model of village development for the Dokra artisans, to turn this traditional craft form into their primary source of sustainable livelihood. In 2013, Contact Base started work with an assessment of the gaps and needs of the artisans and strategized the local interventions accordingly. After the community was mobilized to engage in the process of revitalizing Dokra for a local creative enterprise, interventions focused on capacity development in design and entrepreneurship, technology support through fuel-efficient, minimum-pollution furnaces, upgradation of the village space, marketing and promotion of authentic Dokra craft and the makers, and generally organizing the community into a Dokra producer collective. Equal participation of the community in the process of change led to a more sustained process of development. The enablement of direct linkages with diverse markets led to increased demand for authentic Dokra, appreciation for the artisan community, and cultural awareness about Bikna village. Simultaneously, village improvement was catalyzed through infrastructure development, capacity building of the artisans on offering village tourism, especially through community-led festivals, the building of a community-owned and managed resource center with a workshed, boarding, and lodging facilities for visitors, and a community museum showcasing the history, traditions, process, and accolades of the community.

Bikna, previously an unknown village, has become a hub of enterprising Dokra artisans engaging in regular business, order fulfilment, market outreach, and promotional activities. The craftspersons continuously innovate to produce contemporary designs and products for the changing markets, which attract new buyers. The village itself, which was promoted as a unique cultural destination, also pulls local, regional, domestic, and international visitors on a regular basis. Public awareness and recognition of their specialized skills have also bolstered their collective identity and dignity in their work. Bikna has become a bustling center of Dokra artisan entrepreneurs, with every household busy with their orders and clients. They aspire to grow further in scale and export markets. The skills of the youth have advanced in terms of mixing traditional artisan competencies with the application of digital technology, marketing, and management capabilities. This craft-based entrepreneurial approach to economic development has led to changing lifestyles, improved living conditions, village sanitation, better management of collective resources such as water and drainage systems, social inclusion, women's empowerment, children's education, and overall well-being. The artisans' collective at Bikna, which is registered as a society, is working as the fulcrum of this change, supporting the artisans with capital, management of bulk orders, and organizing the annual village festival.

Chau Masks of Charida village: Chau mask-making is a traditional local craft of Charida village in Purulia district of West Bengal. It was originally an allied industry for the popular masked folk dance drama form Chau (Dey 2012); the artisans made masks of various gods, goddesses, demons, local tribal people, heroes of folk legends, animals, birds, and any other character that the Chau dance groups ordered for their shows. The village of Charida is home to more than 100 such families of mask-makers. The art of Chau mask-making started in Charida around 150 years ago, during the rule of King Madan Mohan Singh Deo of Baghmundi. The masks are handcrafted with paper pulp, clay, and cloth, and then painted to create different characters played by the Chau dancers.

Making and selling Chau dance masks brought the artisans a meager income during the performance season, which lasted for around six months. During the rest of the year, they would manage with an irregular daily-wage income from menial jobs and labor work. These mask-makers did not have any official recognition as artisans under the Artisan Card scheme of Govt. of India despite their unique craftsmanship. As the Chau dance tradition itself had been losing its past glory and patronage, their performances reduced and consequently, the market for Chau masks declined as well. Diminishing income and opportunities from this craft led to a decline in the practice and skills of the artisans and a lack of motivation to pursue this traditional occupation. Hence the Chau mask-making industry was facing a slow death.

Interventions by Contact Base in this community opened new avenues for Chau mask-makers. First, the traditional market for Chau masks was revived by revitalizing the traditional Chau dance industry through innovative productions, capacity building, skill upgradation of the community members by the senior master artists from within the community, market creation, and entrepreneurship support. This led to increased demand for Chau masks. Parallel to this, new diversified products in the form of masks crafted for home decor, wall art, etc., also found a new market. The village gained a new identity as the Charida mask-makers' hub. The interventions for mask-makers included capacity building for product diversification, innovative raw material inputs, branding and promotion, and accreditation of the community by the state government as skilled artisans under the Artisan Card scheme. Exposure and creative exchanges were facilitated for them to understand the modern markets and realize the importance of continuous artistic innovation.

Equipped with their new capabilities, the mask-makers took it upon themselves to develop their village. Charida has become a brand of Chau mask-makers who make different types of masks for visiting tourists as well as for the Chau dance performances. Charida has grown into a full-fledged village industry where artist entrepreneurs have further expanded their own work and created employment for local youth. Artisans work from their homes and all the porches of these houses are workshops of mask-making from where they also sell their wares. The main village street is lined on either side with masks of bright colors hung inside and outside every workshop, with benches outside for anyone to sit and watch the artisans at work and buy masks. All the families have experienced an almost two-fold increase in their production, leading to a substantial growth in their monthly incomes. The Chau Mask of Charida has enjoyed a GI tag since 2018, adding to the branding of their unique art form and products.

Jawaja Leather Association of Rajasthan

The Jawaja administrative block in the Ajmer district of Rajasthan is a traditional hub of leathercraft, believed to be 3000 years old. The aridity of the land and severe climatic conditions, coupled with droughts and water scarcity, make it an extremely vulnerable region for sustainable economic activity. The caste system is ingrained in the social fabric of the local communities, wherein the leather craft community is at the lowest rung of the social hierarchy, resulting in their marginalization and deprivation. The craft involves the tanning of leather from the hides of buffaloes that have died from natural causes, processing the leather for hand-crafting, and then making various leather products. The entire process is done by hand by skilled leather artisan families.

However, the Jawaja leathercraft rural enterprise did not exist before a path-breaking social experiment undertaken by the Indian Institute of Management (IIM)—Ahmedabad and National Institute of Design (NID), two of the country's most esteemed institutions (Matthai et al. 1984). The so-called 'Rural University experiment' started in 1975 with education for self-reliance, and then further evolved to create livelihoods of the artisans through their traditional crafts. The project strategy was to connect the artisan community with contemporary disciplines and knowledge institutions to enhance their capacities.

Their model was that of rural education and empowerment of the villagers based on local resources. Initially, a series of livelihood-generating activities were introduced through rural education and training, such as vegetable cultivation, horticulture, animal husbandry, wool and cotton weaving, spinning, rope-making, leather tanning, and leather products. These opportunities and relevant training opened up a number of economic activities for the villagers of Jawaja. The rural communities in the process became more self-reliant and learned to manage their livelihood-related activities like raw material procurement, finances, bank dealings, and marketing.

As the intervention advanced, it focused on enhancing the know-how and design of the leathercraft artisans to create a village creative industry based on local traditional skills. A collaboration between the Jawaja artisans and NID designers led to the first 'Jawaja collections' with a distinct identity of 'Jawaja' leathercraft and a potential for the Jawaja leather artisans to become owners of their own leathercraft business. Hence, the artisans were organized into a collective, and an artisan organization called the Artisans' Alliance of Jawaja was established.

Initiated in 1975, the Jawaja Leather Association[2] (JLA) is a constituent association of the umbrella organization Artisans' Alliance of Jawaja (AAJ), which continues to be managed independently by the rural artisans themselves through a democratic system of governance. The entity has been a source of sustainable income and livelihood opportunities for all its artisans. It has reinstated a sense of dignity among the artisans and has instilled in them a spirit of self-reliance. Today, JLA is known nationally and internationally for its unique product line. In fact, the Jawaja range contains a classic collection, which has been in demand for over three decades, proving to be an extraordinary example of sustainable design.

To cater to more contemporary markets, these leather artisans have been collaborating with various national and international designers, design students, and researchers from across the world. The economic success of Jawaja leathercraft led to the social empowerment of the community, further inspiring the younger generations to return to their traditional craft. The initiative has improved the overall quality of life in the region. The AAJ has a small common facility center (CFC) in their village where they stock their products, process raw materials, and carry out business activities such as client meetings, quality checks, packaging, and other operational work for marketing. The tanning initially used to happen inside their villages but has now been centralized in terms of stocking up rawhide, tanning of hides, quality checking, and storage.

The CFC has separate tanks for leather tanning and processing, and individual artisans use the facility as per their requirements. They have a local supplier for the hides. These hides are tanned

Figure 4.1 Late Shri Harlal, a leather artisan from Jawaja working in his village home.

Source: JLA.

at the common facility center with water, natural ingredients, and some chemicals. Water is used efficiently as wastage of such a scarce resource is not affordable for the community. It is interesting to note that during the introduction of design know-how by NID, the technical agency on leather in India, Central Leather Research Institute (CLRI), was also engaged to simplify the chemical processing of the leather. The chemical processing system introduced by CLRI, about 30 years ago, significantly reduced the time taken for tanning from six months to one month. This process upgrade has not only made their task less time-consuming but has reduced the drudgery in the leather tanning process by getting rid of the associated odor. Once the tanning is done, the artisans carry the leather to their homes where they use simple tools to make hand-stitched leather accessories. Azo-free dyes are used to add some basic colors to the products.

JLA, the association under AAJ, is not only an authentic hand-made leathercraft rural cluster but its products have also gained the tag of green products, the initiatives of which actually started quite early in their journey. The life history of JLA is unique in that it started not as a craft project but rather as an effort toward self-reliance and empowerment in a highly discriminatory society. When the cluster showed potential to become a role model for the handcrafted leather industry, appropriate inputs were given by specialized organizations to help it turn into a successful enterprise. AAJ (registered as a cooperative) has been functional since 1975. The use of local resources, integrity of values and work, unified nature of the collective, consciousness of the effects of their production on the environment, sustainability of design, etc., are some of the key indicators of the success of this cluster. They are very conscious of the quality of their products and take pride in the fact that their products' longevity, when used regularly, is as much as 20 years. They maintain their high quality by not compromising on the tedious production process and have successfully created a high brand value of Jawaja leather products, especially bags. Exports, which comprise a substantial part of their business, are done directly with the buyers but are facilitated by local buying agents who take care of all shipping-related paperwork, testing, export compliance, etc. Their relationships with the buyers are strong, based on trust and appreciation, with most of them working with Jawaja for more than a decade. Although their business size has decreased over time when some of the international buyers have reduced their business scale, the artisans continue to operate on their own terms, not compromising on quality and relationships.

Bhujodi weavers' village of Gujarat

Bhujodi is a traditional weavers' village in the Bhuj region of Kachchh district, Gujarat. The Kachchh desert is home to an extraordinary variety of artisanal products, which are integrally linked to the community lifestyles, and have their genesis embedded in trade, agriculture, and pastoralism of the region. While Kachchh is often called the handicrafts capital of the country, Bhujodi is one of the mainstays of traditional weaving of 'Kachchhi' shawls, traditional blankets, and stoles.

The weaver communities belong to the Marwada subclan of Meghwals, locally known as Vankars, who are believed to have migrated from Rajasthan about 500 years ago and settled in this region with the engagement of weaving woollen veil cloths and coarse woollen blankets for the Rabari community. Traditionally, the Rabaris are nomadic shepherds herding goats and sheep, and needed woollen shawls and blankets to survive the harsh cold of Kachchh winters. The Rabaris thus provided hand-spun woollen yarns to the weavers to be turned into body wraps and traditional dress fabrics for them. The woven designs, which suited the requirements of the Rabari community, eventually created an identity for these weavers, making Bhujodi village famous for its exquisitely woven traditional textiles. Traditionally, the weavers also used locally grown coarse cotton to make clothes for local use.

With time, the unique thread work, color aesthetics, and designs of the Bhujodi weavers became extremely popular among general consumers, beyond the local indigenous communities who originally engaged in barter trade. Eventually, this local weaving came to be known as Bhujodi weaving, taking its name from the village. Every Bhujodi household has more than two looms, catering to local markets or outside markets and urban consumers.

However, changing contexts and consumers have also led to shifts in the traditional practices of the weavers. During the 1960s, as cheaper mill-made cloth and factory-made products became widely available, the local markets of these weavers declined. Their production process was disrupted, and they were forced to look for external clients and shift their practices to fit the demands of larger urban markets. Hence, to keep up with outside demands the weavers started using synthetic and imported fiber which are easier to weave than the local traditional coarse cotton. This changed their traditional skill and uniqueness to some extent.

After the 2001 earthquake in Gujarat, there was rapid industrialization in Kachchh. Although this brought new prosperity, it also adversely impacted the traditional livelihoods of the local communities. For example, the number of weavers in Kachchh declined from over two thousand in the mid-1990s to only a few hundred. The lack of an enabling environment in terms of access to raw materials, and the inability of small-scale weavers to deal with changing and new markets further necessitated the development of a local value chain for making the weaver community self-sufficient and insulated from external market fluctuations.

In response to these challenges, Khamir,[3] a Kachchh based NGO, began its initiative of developing an indigenous cotton-based local supply chain to revive both cotton farming and traditional cotton weaving while targeting modern aesthetics. Khamir was set up in 2005, with a mission to promote the traditional handicrafts of the Kachchh region to protect and sustain not only the local culture but also the practitioner communities and their local environments.

This local species of cotton is called Kala cotton. Kala cotton is an indigenous organic cotton grown in Kachchh without any use of pesticides and synthetic fertilizers and requires low investments. It is an old-world cotton, which farmers and weavers worked with together in the traditional market systems to create organic woven textiles with a soft but durable texture. Unfortunately, with the introduction of imported new-world cotton by the British and more recently genetically modified Bt-cotton, which could be used to produce finer cotton, the use and significance of Kala cotton had declined. Thus the unique indigenous value chain involving the farmers, weavers, dyers, and markets had been broken, leading to the disintegration of the local economy.

One of the most well-known case studies of Khamir is that of Kala cotton, which led to the revival of sustainable cotton textile production, and the preservation of agricultural and artisan livelihoods in Kachchh (Jha 2013). In 2007, Khamir partnered with Satvik, an association of organic farmers in Kachchh, to explore the production possibilities for Kala cotton with the aim of establishing a local, natural fiber as an alternative to synthetic fiber for the weavers. The great advantage of this indigenous species of cotton is that the crop is purely rain-fed and has a high tolerance for both disease and pests. Hence, it is resilient to difficult land and climatic conditions. The disadvantage was that the cotton's short staple length made it difficult to spin and weave.

Khamir and Satvik first consulted with many experts to develop a process for converting the cotton into yarn. Additionally, local weavers had to be convinced of the value of Kala cotton—a challenge, as weaving it required changes in the loom set-up, as well as differing yields and shafts. Years of R&D led to the establishment of spinning and weaving techniques that could be used to create woven products for modern-day consumers. Their initiative led to the development of a local supply chain between the Kala cotton farmers, ginners, spinners, and weavers to convert raw cotton into handwoven products of a strong, coarse, stretchable fiber. As Khamir's organizational work also involved fostering entrepreneurship among the local handicraft producers, they supported the weavers with business development, training on the use of various kinds of yarns to promote diversity, yarn dyeing, design development, and market linkage facilitation through exhibitions and promotion. Khamir positioned Kala cotton fabric as a sustainable fabric which is handwoven and reached out to designers for further development of the woven fabrics. Many designers came forward to work with Kala cotton weavers, and the community also found designer patrons who contributed to their unique lines of products, designs, and fashion wears.

Khamir began producing the first Kala cotton goods in 2010 by successfully innovating methods to weave Kala cotton into marketable products. A sustainable local value chain was thus established in harmony with the local ecology, as well as by empowering local marginalized communities.

Many of the enterprising weavers have started using Kala cotton for their woven textiles. Kala cotton is also being cultivated today on a larger scale because of the renewed demand. Khamir began working with eleven Kala cotton farmers and also tried connecting them to mills to create a steady demand for this cotton. More than 100 farmers are now practicing Kala cotton farming. There are about 250 families living in Bhujodi village engaged in weaving, and many of them use Kala cotton for their unique repertoire.

Kala cotton already had an organic certification from earlier, which had been facilitated by a local network, Kutch Nav Nirman Abhiyan, with support from Satvik. The new demand along with the existing organic certification enabled premium value creation of Kala cotton in the markets. According to local organizations, the price of Kala cotton went up from INR 700 per 40 kg to INR 1500–1600 per 40 kg.

A group of four enterprising weavers created the Bhujodi Weavers Cooperative which brought together all the weavers in Bhujodi, strengthening their collective identity and negotiation power. Now Bhujodi is a popular tourist destination for weaving, although the weavers are spread all over Kachchh. Bhujodi receives more than 15,000 tourists during the tourist season from November to February, according to the local weavers.

The local weavers also set up a Kachchh Weavers' Association and successfully applied for a GI tag for the Kachchh shawl, strengthening their common brand *vis-a-vis* powerloom products.

Figure 4.2 Workshop of Vankar Vishram Valji in Bhujodi.

Source: Khamir.

The most famous and largest Bhujodi weaving enterprise is that of Shri Vishram Valji Vankar, who won India's prestigious National Award for weaving in 1974. This gave confidence to the family to pursue their traditional craft of weaving and his six sons joined his business. They started working with a wider variety of yarns, such as wild silks of *tussar*, *eri*, and *muga*. They also mixed different fibers to produce fabrics of different textures and look— combinations of wools, silks, and cottons to achieve new kinds of fabrics that can be woven on traditional looms and can be dyed with natural colors. Vankar Vishram Valji Weaving is an iconic example from Bhujodi of a multi-generational craft initiative for weaving and dyeing of fabrics, shawls, stoles, furnishings, and rugs in cotton and indigenous wool, employing around ninety families of this village. Shamji bhai, one of the sons of Vishram Valji Vankar who started leading the family business, was also one of the early users of Kala cotton.

Today the local spinners and weavers work with Kala cotton and the local industry has been regenerated. There is improved awareness among the weavers about sustainability, the importance of using natural materials, the value of maintaining indigenous designs, and the market trends that enable them to create exclusive and modern product ranges. Their ability to interact with markets, consumers, and designers has provided them with exposure and ideas for innovations and creativity.

The narrative of Kala cotton has specific relevance for its contribution towards the generation of a self-sustaining local economy where the cultivation of the indigenous crop and its conversion to products happen within a limited geographical area of about 250 km. The initiative has attracted many collaborative projects and interventions by designers, NGOs, etc., who have promoted Kala cotton initiatives and products.

In 2014, Shamji bhai participated in the 'Hand Made' project exhibited in Bunka Gakuen University in Japan and presented the traditional textiles produced in Bhujodi, including Kala cotton. In 2015, he participated in the project 'Cotton Exchange: A Material Response,' which delved into the social, cultural, and historical legacies of cotton manufacture and trade between England and India, highlighting the case study of Kala cotton. These international collaborations heightened the global awareness of the potential of Kala cotton textiles. A recipient of several awards,

Shamji bhai has led the way for the Bhujodi weavers towards a sustainable approach to production including the conservation of heritage, skills, ecology, and traditional knowledge. Today, with growing consciousness among consumers about buying and using sustainable products, Kala cotton weaves and fabrics have become one of the most popular, sought-after products, being sold by multiple e-commerce platforms, and by designers who are creating high-end collections with the Bhujodi weavers. Bhujodi has become a well-known active village hub of traditional weavers whose products are popular across the world.

Conclusion

The case studies discussed in this chapter showcase how local rural communities across diverse cultural and social backgrounds have collectively contributed towards strengthening their village economies based on their traditional creative skills and resources. They can each be viewed as a 'cultural village,' which is typically defined as a certain geographical area that has historically borne a strong cultural identity and skills. The village or the area is therefore potentially home to cultural and creative industries which thrive at the intersection of culture and socio-economic development. Each case demonstrates the importance of inclusive, participatory, collaborative, and cooperative decision-making, reaffirming the conclusion of Permana and Harsanto (2021) that such culturally-led approaches have the potential to successfully revitalize traditional environments.

The models discussed also demonstrate the key principles of human development theories and rural redevelopment strategies, as well as the ideas and ideologies postulated by economists and social reformers, discussed in Chapter 2. Rather than trying to achieve the neoclassical notion of economic development based on industrialization, the development approaches used in the case studies address the large but traditional and underdeveloped rural subsector of India, which still represents the majority of its population, and is characterized by widespread poverty, unemployment, and low productivity. In the alternative human development theories in the field of development economics, a people-centric approach is cardinal, which is also the key to the successes and sustainability of these case studies. The human development approach focuses primarily on enlarging and expanding people's choices and providing them the opportunities they need to lead healthy and meaningful lives, supported by education and a decent standard of living, ensuring human rights, self-respect, and political freedom. Stewart, Ranis, and Samman (2018) describe this as a two-way approach to maximizing societal happiness along with the pursuit of economic growth. To achieve improved human development while sustaining economic growth, it must be combined with societal well-being, equity, social capabilities, and the active role of social norms and organizations based on collective ideology and action, *vis-a-vis* individualistic approaches.

Echoes of these ideas are found in the works of Mahatma Gandhi and Rabindranath Tagore as well. Tagore believed in a parallel economy of small and cottage businesses which are as relevant as large-scale technology-based industries, especially for achieving holistic human and societal well-being based on cooperation, leading to balanced growth of both cities and villages (Dasgupta 1993). Gandhi believed in economic decentralization and considered villages as the basic economic unit. His development framework of Gram Swaraj has a correlation with the case studies in terms of a decentralized, simple, village-based small-scale economy, where means of production are local and accessible to the rural producers, and where social welfare is attained through dignity of work, enhanced community spirit through local economy, reduced marginalization, and less adverse exploitative conditions (Kakati 2021).

The case studies discuss how the local traditional cultural and creative skills of rural communities have been instrumental in building the capabilities of the villagers to acquire opportunities for a healthy decent livelihood, self-respect, and overall community development. The transformative outputs described above illustrate how culture and creativity provide for a personal and community identity and develop a foundation for human rights and freedom of expression.

It is important to reiterate here that the concept of self-sufficiency used in these case studies in no way suggests an isolated economy without any contact with the outside world. The idea of the isolation and self-sufficiency of the traditional Indian village was first propounded by Sir Charles Metcalfe in 1830 and took root in subsequent sociological and anthropological literature on

Indian villages. Srinivas and Shah (1960) have criticized the idea of self-sufficiency and isolation of Indian villages as illusory, possibly stemming from the difficulties in accessing the villages owing to bad roads and connectivity, heavy monsoon rains, the existence of barter systems, etc. Beteille (1966) in the context of his Sripuram studies also reiterated that the village was never fully self-sufficient in the economic space, as far as living memory went. The regional socio-political and economic histories discussed in Chapter 3 also clearly suggest that Indian villages had been part of wider economic networks from ancient times.

In the modern era, external organizations such as NGOs and educational institutions have always played a key role in the community empowerment processes. In each of these cases, the idea of rural development in terms of improving the economic and social well-being of a specific group of rural people who are poor has been undertaken by external organizations, which have played a critical role in mobilizing, capacity building, facilitating, and handholding the development of the village economies. However, the process of change has been bottom-up to a large extent, by making the local communities an integral part of planning, decision making, and leading their own process of empowerment. Although this inherent ownership of these initiatives by the local communities has been critical to the sustainability of these interventions, the commitment of the external organizations in providing handholding support to the rural creative collectives as and when necessary, has also been a guiding factor encouraging the rural change-makers to continue empowering their fellow villagers, thus strengthening the self-sufficiency and resilience of the communities.

Notes

1 https://ruralcrafthub.com/ (accessed May 2023).
2 www.jawajaleather.com/ (accessed May 2023).
3 www.khamir.org/ (accessed May 2023).

References

Bajpai, Lopamudra Maitra. 2015. "Intangible Heritage Transformations – Patachitra of Bengal Exploring Modern New Media." *International Journal of History and Cultural Studies* 1 (1): 1–13.

Beteille, Andre. 1966. *Caste, Class and Power: Changing Patterns of Stratification in a Tanjore Village*. Oxford University Press, New Delhi, India.

Bhattacharya, Ananya, and Madhura Dutta. 2022. "Empowering Heritage Entrepreneurs: An Experience in Strategic Marketing." *Journal of Heritage Management* 7 (2): 186–99.

Chakroborty, Ratnadip, Soumita Chatterjee, and Sutapa Choudhury. 2021. "Changing Paradigm of Life: An Empirical Study Among the Dokra Brass Casters in West Bengal, India." *Journal of Research in Humanities and Social Science* 9 (2): 17–29.

Dasgupta, Tapati. 1993. *Social Thought of Rabindranath Tagore: A Historical Analysis*. Abhinav Publications, India.

Dey, Falguni. 2012. "Folk Culture of West Bengal." *Journal of Institute of Landscape Ecology and Ekistics* 35 (1): 537–550.

Jha, Banhi. 2013. "Kala Cotton: A Sustainable Alternative." In *The Asian Conference on Sustainability, Energy and the Environment 2018. Official Conference Proceedings,* The International Academic Forum, Nagoya, Japan.

Kakati, Bhaskar Kumar. 2021. "Gram Swaraj: Its Relevance in Present Context." www.mkgandhi.org/articles/articleindex.htm.

Matthai, Ravi J., Helena Perheentupa, Nilam Iyer, and Ravinder Kaur. 1984. "Learning for Development at Jawaja." *India International Centre Quarterly* 11 (4): 105–11. www.jstor.org/stable/23001709.

Permana, Chrisna T, and Budi Harsanto. 2021. "Decision Making in the Culture and Creative Industries Environment: Lessons from the Cultural Village." *AFEBI Management and Business Review* 6 (1): 1–11.

Srinivas, Mysore Narasimhachar, and Arvind M Shah. 1960. "The Myth of Self-Sufficiency of the Indian Village." *Economic Weekly* 12 (37): 1375–78.

Stewart, Frances, Gustav Ranis, and Emma Samman. 2018. *Advancing Human Development: Theory and Practice*. Oxford University Press, New York.

5   Rural creative entrepreneurship

The active cultural and creative hubs such as the ones described in the previous chapter owe their success in part to individual artisan entrepreneurs from within the rural communities who champion the local transformation towards an alternative economic, social, and cultural model of growth. The leadership of such individual artisan entrepreneurs from different generations has provided a significant boost to the development of rural creative entrepreneurship across different crafts and communities. In this chapter, we look at the stories of a few such entrepreneurs.

Tajkira Begum, West Bengal

Tajkira is a successful rural entrepreneur in Kantha embroidery, which is a traditional craft of Bengal. Kantha is the art of sewing together pieces of old fabric to create a new fabric which is further ornamented with surface designs of fine threadwork. The traditional product is a quilt mainly used for covering oneself or children. It had been a popular utility item of every Bengali household, and the craft was practiced by rural women of all classes.

Kantha as a craft, with beautiful traditional motifs and patterns inspired by the women's daily life and backyard biodiversity, also became popular among designers for creating contemporary collections for high-end markets. However, the rural Kantha artisans seldom received a fair share of the market returns, as they lacked the capabilities and confidence to deal with modern buyers directly.

DOI: 10.4324/9781003331476-5

Tajkira lives in Nanoor, a village in the Birbhum district of West Bengal. It used to be an ordinary village of a socially marginalized community, living largely in poverty. Its residents, most belonging to Muslim and other backward communities, mainly engaged in daily labor jobs. Some women in this village retained the skill of Kantha making and also did some daily job work for local customers, but never thought of turning it into their business. Tajkira Begum, among them, was an enterprising woman who dreamt of building her own craft business in Kantha work. She used to travel to nearby towns to bring job-work orders and deliver finished products in exchange for meager wages. She toiled to support her family economically and otherwise. Being a woman and a Muslim, her free mobility to towns was not acceptable to many villagers, who questioned her dignity and intentions openly. In spite of such hindrances, Tajkira continued her journey.

When Contact Base intervened with an entrepreneurship support program in Nanoor, Tajkira actively joined the initiative (Palit and Debnath 2022). Through various trainings she honed her skills, started making diversified products and innovating her designs, and gained knowledge on business development and management. She got exposure through her visits to exhibitions in the big cities and at national and international levels with support from Contact Base. Soon, she established her own market linkages and started bringing orders from diverse markets and customers. She was not a job worker anymore but started her own independent business. As the order and client base increased, Tajkira herself trained fellow women of her village and motivated them to join her. She would generate the business and distribute work to more and more village women, providing them with work at home and empowering them economically. The direct business linkages ensured more earnings and profits for Tajkira which she invested to further expand and strengthen her collective of artisans for catering to larger markets. In Tajkira's own words, this was like a dream come true in a village where women were previously confined to their homes and had no source of independent income except for some local job work with paltry earnings of INR 500–600 per month. Hardly any girl children in the village used to go to school, and boys' attendance was also sparse. With the economic independence of the women, their lives changed, and their voices grew stronger. They could send their children, especially girl children, to school, and could make their own choices in the family. In a village where there were once only a few who completed secondary education, and did not think of higher studies, there are now almost 20 women who have completed their Masters. They see education as a necessity, not an option. Over time, the women's dignity of work has been established leading to pride and recognition of the entire village.

When Tajkira initially expanded her business, she distributed work to more than 300 women of Nanoor. In the process, she assessed their skills, self-motivation and interest, and sense of responsibility and accountability for future planning. She also tried her best to pay the women workers on time even when clients delayed payment. This helped her to motivate and sustain the engagement of these women and build trust and long-term relationships. She realized that strengthening the collective was fundamental to leading her Kantha business. She also identified women on her team who showed entrepreneurial flair and trained them to become entrepreneurs. Under her leadership, more women entrepreneurs started emerging, gradually extending the work to include even more village women. Under Tajkira's leadership, they understood that unity, collective identity, and lifting each other up are key to success in building their sustainable businesses.

Out of the 300 women she started working with, 100 have become small entrepreneurs with new capacities learned on the job under Tajkira. These new entrepreneurs have further expanded to provide work to a cluster of 600 women engaged in Kantha embroidery work professionally. Tajkira herself enjoys a good sustainable livelihood and has been the epitome of success for the entire village.

Awdhesh Kumar, Rajasthan

Hand block printing on fabrics is an age-old tradition of India with a wide regional variety practiced by different communities in different parts of the country. One of the famous traditional centers of this craft is Sanganer in Rajasthan, where the Sanganeri hand block printing technique originated. Sanganer, a suburban administrative subdivision in Jaipur district, is well known as a tourism destination, being home to a large number of handicraft artisans. Sanganeri textile printing is believed to have developed between the 16th and 17th centuries. As a brand, it enjoyed prominence facilitated by rich traders and royal families, was a major export item for the British East India Company and continues to be important in the modern era.

The artisans belong to the Chhipa (literally meaning 'printer') community who are Hindus and are believed to have migrated mostly from Gujarat. An artisan's entire family gets involved in the textile printing process. Specialized sub-industries of this craft include printing, dyeing, and block-making. The block carver's community of Sanganer are small traditional businesses that work on a typical design repertoire, innovate new designs as per orders, and also mend old blocks that have been worn out from repetitive use.

One of the iconic Chhipa families, traditionally engaged in Sanganeri block printing craft through five generations of artisan entrepreneurs, is that of Awdhesh Kumar, which has turned into a brand name today. In the 1940s, Harnath and his son Balchand Pandey had started a small local business in hand block printing producing dupattas and fabrics, which they sold in a road-side local market. In their first business itself, they created a brand name Balji Harnath Hala which they put on each of their products to assure their customers of the uniqueness and authenticity of their products.

Ram Kishore Pandey, of the next generation, started a company in 1971 by the name of Ram Kishore Awdhesh Kumar to sell their products to a larger clientele of government emporiums and other companies. His son Awdhesh Pandey continued their legacy by starting two separate brands in 1994, A. K. International dealing with exports, and A. K. Textiles for domestic markets. He served as the Managing Director of the AK Group. Awdhesh Pandey is recognized as a master artisan, a block printer, a craft revivalist, an expert in natural dyeing, and a teacher in hand block printing. He has taught about 2000 students from India and abroad since the 1990s, not only in his workshop but also as visiting faculty in national design and technical institutes. He received many prestigious national awards for his contribution to their traditional Sanganeri hand block printing craft.

Awdhesh Pandey's younger son Khushiram Pandey went to formally study design at a renowned design institute in Jaipur. Entrepreneurship was an obvious choice for him too. After

Figure 5.1 Khushiram Pandey (right) at work in his hand block printing unit.

Credit: Khushiram Pandey.

completing his education, he started the brand Awdhesh Kumar in 2012,[1] focusing on expanding the retail business. His specialization in textile product design, and his final year project on creating a collection of prints and garments inspired by regional architecture and the City Palace Museum of Jaipur made him ready for innovation and product development for modern tastes and demands. Khushiram is currently the Creative Director of AK Group and also works with his own brand which deals with exports, wholesale, and retail businesses.

Theirs is an exemplary case study of an artisans-led creative industry based on high business acumen, business innovation, respect for their family tradition, and a continued belief in the potential of their craft. The vision of this family has been to revive and continuously revitalize the ancestral work of hand block printing and natural dyeing, skills that have been passed on and preserved through generations in the family. With time, and in response to the needs of changing markets and consumers, they have diversified their work by combining modern aesthetics with different types of traditional printing techniques. They have innovated not only with new product lines but also with different types of base materials such as cotton, linen, and silk. The enterprise now works with about 100 artisans providing them with regular work and fair wages. They have also built their own library of almost ten thousand wooden blocks of traditional and contemporary designs.

Adil Mustak Khatri, Gujarat

Bandhani is a tie dye technique practiced in Gujarat and Rajasthan, Kachchh being one of the best-known centers of its production. It is believed that the technique was brought to Kachchh by craftspersons of the Khatri community who came from Sindh, probably in the 16th century. Traditionally Khatris made Bandhani only for themselves but today it is a highly sought-after craft across the world, with its large repertoire. Bandhani has a robust, wide-ranging market and provides income for the traditional artisan communities who pass down the skills through generations.

Adil Khatri is a young Bandhani artisan and entrepreneur who belongs to a traditional Bandhani artisan family.[2] Adil was born in a middle-class family of traditional Bandhani artisans of Bhuj in Kachchh. Women artisans are traditionally involved in the tying process in creating Bandhani tie and dye, while men do the dyeing. Adil learnt the intricate craft of tying from his mother and dyeing from his uncle, and he chose to become a full-time Bandhani artisan after completing his higher secondary education. By creating new products with his craft and design skills, Adil wanted to preserve his family's tradition through entrepreneurship. With his aspirations to start a successful business, he went to study design and business management in Somaiya Kala Vidya (an institution for design and business education for traditional artisans of Kachchh) in 2014. After completing his education, he launched his own brand, Nilak,[3] which has made a mark of its own among the numerous Bandhani craftspersons of Kachchh. His innovative designs and uniquely attractive color palettes have made his work special. Adil won many awards for his excellence in Bandhani craft and his success as a young entrepreneur, especially at a time when youth from craft communities are increasingly leaving their traditional occupation and skills for urban jobs in cities.

Figure 5.2 Adil dyeing.

Credit: Adil Mustak Khatri.

Every year, he presents his work at national and international forums, exhibitions, and workshops with the aim of demonstrating his crafts and sustaining his livelihood. He also appeared on India's renowned news channel NDTV's show *Icons of India*, where he explained his craft techniques. In 2021, he and his partner, a fellow Bandhani artisan, participated in the International Folk Art Market in Santa Fe, New Mexico, which has been one of the most premium international art and craft markets for decades.

Conclusion

This chapter presented three case studies, from three different geographies and socio-cultural backgrounds, of rural creative entrepreneurs. One can find many similar success stories across India. Even though factors such as caste, gender, and other social divisions have been a hindrance to their entrepreneurial endeavors, their conviction and grit, along with their traditional skills and strong cultural rootedness have enabled them to emerge as successful creative entrepreneurs. As traditionally handicrafts industries of India are community-based, the role of the village community artisans in promoting their traditional crafts has been important in building the village creative enterprises, eventually employing more village people in production and business, thus strengthening the collective as a whole.

These case studies are pivoted on the concepts of 'agency,' 'capabilities and functioning,' and 'empowerment.' In the field of development economics, Amartya Sen pioneered the concepts of agency and freedom of choice to explain poverty (Sen 1988). Sen put forward that the capability of people is directly linked to their reasoned agency to function in a way that brings about positive change in their own lives. Capabilities refer to the set of valuable functionings that enable a person to choose particular aspects of life in terms of their physiological, emotional, and physical environments, and decide to utilize their resources for their subjective well-being.

This conceptual framework has been subsequently expanded and broadened by other theorists. In particular, the idea of capability was further elaborated by Martha Nussbaum, who developed a more concrete framework of the capabilities approach inspired by the Aristotelian and Marxian ideas of human flourishing and good life (Nussbaum 2000; Pruitt 2010). She emphasized that an individual's well-being depends on their life activities based on rational choice, and lists ten essential human capabilities, termed as central human capabilities without which no human being can lead a flourishing and dignified life (Saigaran, Karupiah, and Gopal 2015). She combined the central human capabilities and created a 'threshold of capabilities' that needs to be guaranteed to every human being at par with other social minimum securities and infrastructures. She also believed that government and other social institutions have a role to play in ensuring the provision of these core capabilities. Both Sen and Nussbaum establish the importance of human capability (the ability to achieve) and functioning (achievements) of capable humans to empower themselves to improve living conditions related to social inequality, gender, education, health, and other deprivations.

Development theorists have also adopted these concepts as powerful conditions for poverty alleviation and individual as well as community empowerment. These ideas also inform practitioners' approaches and strategies for catalyzing such processes of positive transformation through human capabilities and human agency.

Narayan (2002) describes a conceptual framework for understanding empowerment as increasing and improving an individual or group's capacity to enjoy freedom of choice and to act to transform those choices for the betterment of their own lives. The importance of an opportunity structure is emphasized for the effective use of agency successfully, where the opportunity structure embodies institutional, social, and political conditions, associated rules and norms within which the members of a society act, and individual (physical, physiological, psychological, emotional), as well as collective assets (community voice and identity).

The case studies discussed in this chapter can be viewed as practical demonstrations of generating capability, agency, and functioning of individuals who aspired, made rational choices, and acted for achieving a better future and well-being for themselves, and their community at large.

Notes

1 https://awdheshkumar.in/ (accessed May 2023).
2 www.facebook.com/adilkhatri, www.instagram.com/adilmkhatri (accessed May 2023).
3 www.facebook.com/nilakbyadil/ (accessed May 2023).

References

Narayan, Deepa. 2002. *Empowerment and Poverty Reduction: A Sourcebook.* World Bank, Washington, DC.

Nussbaum, Martha C. 2000. *Women and Human Development: The Capabilities Approach.* Cambridge University Press, Cambridge, UK.

Palit, Sreenanda, and Debalina Debnath. 2022. "Women as Game Changer of Rural Economy: A Case Study of Kantha Artisans of Nanoor." In *Women Empowerment in India: The Changing Scenario,* edited by Krishnendu Roy, 223–35. Red'Shine Publication, Stockholm.

Pruitt, Lisa R. 2010. "Human Rights and Development for India's Rural Remnant: A Capabilities-Based Assessment." *UC Davis Law Review* 44 (3): 803–57.

Saigaran, Nithiya Guna, Premalatha Karupiah, and Parthiban S Gopal. 2015. "The Capability Approach: Comparing Amartya Sen and Martha Nussbaum." In *Proceedings of Universiti Sains Malaysia International Conference on Social Sciences 2015.* School of Social Sciences, USM, Malaysia.

Sen, Amartya. 1988. "Freedom of Choice: Concept and Content." *European Economic Review* 32: 269–94.

6 Urban enterprises for traditional craft-based creative industries

A major distinguishing factor for Indian handicrafts has been their extraordinary nature of being almost entirely handcrafted, or handmade. The increasing outreach of diverse markets to the Indian handicrafts sector has created new avenues of business and income for the rural artisans. However, this has also brought challenges posed by the mechanized production of identical craft products. Handloom designs are copied and replicated in powerlooms, hand block prints are replicated in screen prints, handmade processes are transferred to mechanized processes making production easy, low-cost, and scalable. Such products are often sold as handmade by fraudulent market players to maximize their profits. This practice has had a substantial impact on the market for authentic handmade crafts because consumer awareness about the authentic handicraft traditions has been low. There are several reasons for this: the vastness, diversity, and remoteness of the practitioner villages in rural India; lack of documentation and promotion of the craft traditions and their makers; and lack of capacity of the rural artisans to establish their authenticity, rights, and value in the market.

Creating differentiation for handcrafted products leading to higher value and market share has always been a key marketing strategy for handicrafts. While a consumer may choose to buy a mechanized copy of a traditional handicraft, and indeed there are such markets that operate legitimately, the fraudulent practice of selling a mechanized product as a handcrafted product, with increased pricing, has been rampant in the Indian market ecosystem. Although progress in terms of effective policies to counter

DOI: 10.4324/9781003331476-6

this phenomenon has been slow, a strong business ecosystem has developed, driven by urban entrepreneurs, with the aim of promoting and selling authentic handicrafts sourced directly from the practitioner communities and rural artisan-based enterprises. These private enterprises have strengthened and grown over the years, showcasing the exquisiteness of India's handicrafts and also creating value by promoting the highly skilled hand processes and indigenous knowledge of the makers.

The case studies in this chapter will present the innovative and enterprising models of some of these urban enterprises and brands that created aspirational buyers from all across the world for high-end authentic Indian handcrafted products. They created a business ecosystem which included rural artisans and entrepreneurs, contemporary designers, sector specialists, and product innovators, supported by strategic branding, consumer education, visibility creation, and market development.

Fabindia

Fabindia is a for-profit Indian retail company which started with the mission of marketing handmade artisanal products of India directly sourced from the makers.[1] It was established in the 1960s as a company that worked with various craftspersons and weavers across the country and developed home furnishings that were exported to retail stores in western markets. Fabindia played the role of an intermediary, buying the products from artisans and weavers and taking their products to the urban markets worldwide, with the objective of providing them with employment and income. They started their first domestic store in 1975 on an experimental basis, and it was a big hit. By the early 1980s, Fabindia had started a garment line made from hand-woven and hand-printed fabrics. It eventually expanded into a consumer-facing retailer selling various handmade products, for apparel, fashion, and home, through its stores across India and overseas. It is the largest and one of the most well-known retail brands, with over 100 stores, which has made handicrafts aspirational in the Indian domestic market, especially for the general population and youth.

Fabindia has been one of the most highlighted case studies in business management courses for the model it adopted, keeping in mind the community-based nature of handicrafts production and business, and scaling it up (Konwar 2011). Fabindia sourced products from the traditional producer communities in the villages who worked from their home setup and in reality had almost no access to the broader markets. In the process of working with the artisans and weavers, Fabindia also trained them in achieving the necessary quality, design aesthetics, and scale for being successful in the market. To scale up Fabindia, their supply network had to be increased by including many more communities across India. A major challenge was to ensure quality standards of products sourced from different craftspersons of different regions who are making them with their hands, making each piece unique. Although there was the need for a centralized coordination agency to manage production and supply, this was impractical because such work required a thorough understanding of the processes of traditional local crafts from across India's nooks and corners.

Fabindia innovated their model to reorganize the company and create 17 supplier region companies (SRC) covering the entire country. A minimum of 26% stake in each SRC was reserved for artisans, with the idea of making them part owners of the company. These direct partnerships with the craftspersons have been a USP of Fabindia, adding value to their brand; Fabindia owned a 49% stake and employees and investors owned the rest. The shares of these companies could be traded twice a year. Most of these SRCs became profitable and paid dividends. A separate microfinance company was set up to finance these SRCs. Thus, Fabindia management established an orchestrated supply chain, which not only catered to the need for a large volume of supplies but also the desired market-oriented design and quality. SRCs provided artisans with inputs on designs and market trends, facilitated their access to funds and management skills, and ensured product quality for Fabindia. Thus, quality control and sourcing were decentralized and localized. As the SRCs supplied directly to the retail outlets, the time to market was also reduced.

Anokhi

Anokhi is a brand that has been working solely with the hand block printing craft tradition of Rajasthan since its inception in 1970.[2] It started its business with exports. Their aim has been to create beautiful marketable products by blending contemporary sensibilities with traditions of excellence to provide sustained and rewarding work to the actual craftspersons of this tradition. Anokhi started very small as a supplier of high-end, high-quality block printed garments to a boutique in the UK. Along with the designs of the founder herself, Anokhi in its early years had also started getting design input from international fashion designers. It could easily distinguish itself in the market through constant design innovation, creation of new patterns, and use of distinctive color palettes.

Initially, Anokhi worked with the UK store and an Australian client, engaging in high-quality but small-scale production (Kumar 2006). During this time, participation in the block printing industry in Jaipur was completely caste based. The Chhipa caste community practiced this handicraft and worked from their village homes in clusters of Sanganer, Bagru, and Jahota. Every cluster was unique in their types of prints and colors which signified their social status, identities, customs, and specific occasions in which certain prints were worn. Anokhi started training the block printers they worked with on the importance of design innovation. With time, Anokhi diversified to include more product lines such as home furnishing items and accessories. Their model involved investing in the fabric, strengthening the production base by capacity building of the block printers in improving quality and design diversification, promoting their product as 'high-end,' and ensuring consistency in quality of products.

For about two decades, Anokhi remained a purely export company gaining success in the booming international markets of the United States, the United Kingdom, other European countries, and Japan. They turned towards the domestic market in a big way only when they faced stagnation in export sales during 2002. The company started its first domestic retail store in Jaipur, in 1981, where it started selling export surpluses and products with minor defects. In 1988, they opened a second retail store which attracted mainly foreigners and tourists and had a limited customer base. The transition in the 2000s led to Anokhi's investments towards changing designs and products for Indian markets. It was also found that the high-value handcrafted products of Anokhi achieved better sales in India than abroad, which they attribute to a greater awareness of craft traditions among Indians. Over the years, Anokhi started targeting more mainstream markets with a diverse customer base of different demographic profiles, for which it kept innovating designs and products, and added various sizes and shapes more suited to Indian tastes.

Anokhi continues to work directly with the traditional block printers in Sanganer and Bagru to create beautiful fabrics which they turn into products in their in-house facilities. They also have their own product sampling facility with a printing unit and an embroidery unit. The two important departments they have are the quality control department, and the tailoring department, both of which have grown significantly over the years, ensuring their brand quality. Anokhi also ensures that the quality of fabric which they give to the block printers for printing is not compromised. It has long-term relationships with mostly NGO-supported handloom suppliers, who are equipped to meet the quality standards of the brand and successfully deliver orders on time in return for assurance of order commitments for the artisans.

Today, Anokhi has a chain of retail stores strategically located in select cities across India. The design skills and business acumen of the founders, along with their vision and ability to institutionalize production systems and processes, especially with the traditional block printers, have played a key role in the success of Anokhi.

Dastkar

Dastkar was formed in 1981 as a private not-for-profit NGO with a mission to support craftspersons, especially women, to use their own traditional skills to earn a dignified income and achieve economic self-sufficiency.[3] The organization worked towards the promotion and revival of traditional crafts by expanding the outreach of the rural artisans to a wide cross-section of buyers in urban India (Tyabji 2003).

Although rural artisans were the actual bearers of traditional handicraft skills and makers of handicraft products, in reality, they were dependent on middlemen and market agents, designers, and traders who maximized their own profits by directly interfacing with the urban markets, keeping the actual artisans in the background. Dastkar, to address this challenge, provided a direct marketing platform to these artisans and built their capacities to ensure that they could cater to the urban buyers on their own and become self-sufficient in marketing their unique products. Dastkar empowered the artisans with training on market intelligence, business capacities, and innovation suiting changing consumer choices. They also addressed the areas of skill evaluation and upgradation, capacity building for design and product development, access of the craftspersons to market trends and information, credit sources, and other networking assistance through various projects across India in collaboration with artisan collectives and local grassroots NGOs or craft organizations. This helped strengthen the artisan and craft enterprises and entrepreneurs. The first Dastkar Bazaar (market) was held in 1982 with 12 participants. In 1983, inspired by the idea of promoting natural fiber crafts which are environment friendly, as well as crafts using motifs from nature and wildlife, they organized the Dastkar Nature Bazaar. Over the years, Dastkar annual crafts fairs have gained immense popularity as a one-of-its-kind exhibit featuring authentic handicrafts of India by the traditional makers themselves. These fairs, which used to be organized at different locations in Delhi every year, caught the attention of a wide cross-section of people and also helped in creating awareness and appreciation for the diverse and exquisite handicrafts of India. Eventually, in 2012, Dastkar leased a permanent space called Kisaan Haat from Delhi tourism to regularly hold curated thematic craft fairs such as Festival of Lights, Winter Weaves, etc. To make the space more vibrant with a multi-cultural appeal, Dastkar also included folk performing arts and cuisines from various parts of India making the platform celebratory of Indian traditional creative industries. In order to create exposure for Indian artisans and value for their work, they also organized fairs with South Asian handicraft producers (Bangladesh, Pakistan, Sri Lanka) to promote cultural exchange.

A key factor for the value creation of Indian handicrafts is consumer awareness of the process of making the crafts, including local materials used and their specialities. The fact that the artisans have unparalleled skills in producing art and craft out of local materials with their bare hands and rudimentary tools is awe-inspiring for consumers. To highlight this aspect, Dastkar also organized workshops and demonstrations by the artisans during the fairs to instill pride in the makers and consciousness in the buyers. Such a business environment enabled the right valuation and appreciation of authentic Indian handicrafts, recognition of the craftspersons, fair business, and promotion of the artisans leading to their further business expansion.

There are several success stories and case studies of Dastkar, showcasing their transformational role in changing the status quo of rural artisan communities from poor, marginalized, wage laborers and struggling entrepreneurs to independent and proud creative producers. For over 40 years, it has been supporting creative and cultural entrepreneurship development, sustainable design, and various public and private collaborations for strengthening artisan livelihoods.

Good Earth

Good Earth[4] was set up in 1996 by a passionate craft connoisseur and a studio potter with the aim of reviving the traditional craft of village potters of India which was languishing (Kapoor 2017). Good Earth thus ventured into creating a retail platform for bridging the gap between rural artisans and urban consumers through contemporary designs and traditional aesthetics. Good Earth opened its first boutique in 1996 as a luxury retail brand and defined luxury as pure, natural with original design, and preferably handcrafted. The founder's fundamental thought behind the brand was to bring the best of diverse and astounding handicraft legacies of India through timeless designs to the world.

The philosophy behind building this brand was to sustain the immense craft heritage of India economically and culturally by collaborating closely with the actual artisans and knowledge bearers of different craft traditions to develop luxury collections of universal appeal, thus celebrating the extraordinary human skills of the country. Good Earth started with luxury products for the home segment. It started with ceramics, tableware, bed and bath linen, and soon began making a range of wall coverings, cushions, decor items, products for children, a line of 'Kansa' utensils using bell metal alloy, etc., all of which were results of craft collaborations in different parts of India.

Over time, Good Earth has taken the Indian heritage of handicrafts to a global platform in such a way that it is often referred to as an ambassador of the rich Indian craft heritage to the world. It offers not only a wide diversity of crafts but also the stories behind them, triggering a hint of nostalgia and aspiration to collect pieces of timeless legacies.

However, the absence of a supply chain in rural India, where the artisans reside, has been the biggest challenge for Good Earth. The team had to travel from village to village to build their own network of craftspersons, ensuring the quality of raw material supplies, working on expected product qualities, etc. Good Earth had started its journey with a small team of potters, designers, and front-end staff, and has grown to be one of the most sought-after high-end luxury brands of Indian craft traditions. During the period when they started their business, liberalization had opened up global market outreach, increasing disposable income, and the awareness and aspiration for multicultural aesthetics was developing. Hence, their traditional home objects and crockery could benefit from the environment by making headway in western markets. Over two decades, Good Earth has become a premium brand, with multiple stores including an overseas presence. They also launched their Web Boutique in 2013.

Good Earth's strategy was never to change the original appeal of the authentic craft but to build on it to extend a royal luxury appeal to its products. Hence, it is believed to have sustained a consistent identity that has made them so successful. Another important design strategy has always been to add a story to a product or collection, making it more than just an item that the customers take back. These stories are linked to the royal past of India, especially in terms of its ancient trades through the Silk Road and other overseas routes.

Good Earth, uniquely, has taken Indian handicrafts to the level of valuable art. The brand partnered with the Victoria & Albert Museum for an exhibition titled 'The Fabric of India,' which showcased handmade textiles of India from the 3rd century to the present day. In 2014, Good Earth was commissioned to participate in the renovation of Jaipur's Rajmahal Palace, and in 2015, the Oscars team invited Good Earth to share gift bags with the guests at the Academy Awards.

Craftmark

Although the creation of a strong network of handicraft enterprises, businesses, and conscious markets has positively impacted the livelihoods, income, profit, and recognition of the actual producers and their handcrafted products, the challenge posed by fake products in the market continued. The sector collectively felt the need for a mark or a seal that would protect the authenticity of genuine handmade crafts and their producers, and thus be a symbol of trust for the consumers. This led to the emergence of Craftmark, designed and run by a national organization, which was set up as a membership-based network of Indian handicraft producers and enterprises. This organization, the All India Artisans and Craftworkers Welfare Association (AIACA) was established in 2004 as a not-for-profit entity aiming to capacitate and promote the authentic handicraft producers and businesses of India for the global markets.

The Craftmark was one of the first certifications in India developed exclusively for the Indian handicraft industry.[5] Launched in 2006, it was created to promote authentic handmade products produced by artisans and weavers, increase consumer awareness of distinct handicraft traditions, and establish a collective negotiation power for this rural and traditional industry. Registered under the Trademarks Registry, Govt. of India, Craftmark is the only market-led national certification program for genuine Indian handcrafted products produced in a socially responsible manner.

The design of Craftmark is simple and inclusive. AIACA licenses the Craftmark seal to artisan organizations, craft-based businesses, cooperatives, and NGOs for use on their products. The interesting aspect of Craftmark is that it certifies a 'craft process' and not a 'product.' Craftmark develops sector-wide, process-specific standards and norms for labeling a product as handmade. Standard verification tools are used for inspecting the applicant facilities and production processes. In addition to certifying the authentic production process and skills of the actual practitioners, Craftmark certifies that the process is socially responsible, that is, minimum wages are paid to artisans, there is no child labor, and working conditions meet basic minimum standards. Craftmark thus addresses the market requirements of 'premiumness,' ethical production, and occupational health and safety.

When an application is approved, the certified party is eligible to use Craftmark labels (tags) carrying a license number, name of the craft, and place of production along with the name of the licensee. Fabindia, for example, has a Craftmark license for a number of handicraft processes, and the corresponding products carry the Craftmark label. Along with such large-scale business entities, Craftmark is also used by smaller craft entrepreneurs and craft collectives. Craftmark has a dedicated website with information about all active members, creating a common platform of certified Indian handicraft entities / businesses and artisans of great diversity. This website contributes to the promotion of the members, especially in the digital age. The Craftmark license is biannually renewable, with fees based on the annual revenue of the applicant.

Craftmark has evolved over the years and grown organically to cover more than 100 handicraft processes across a wide variety of raw materials from different geographies. Craftmark has a total outreach of more than 1,65,000 artisans across 23 states of India. Many Craftmark members use this brand to market their products abroad and are able to gain better prices and recognition for their crafts and creativity. To enable market vigilance for addressing misuse of the tag, Craftmark also introduced the use of QR codes that direct prospective buyers to the respective member page on the Craftmark portal. If a tag is being misused by a non-member, their page on the Craftmark portal will not be found.

Craftmark also has its own marketing department that tries to facilitate and support small artisanal organizations with order fulfilment, quality check, packaging, exports, regulatory paperwork, bank transfers, etc. It caters to buyers, buying houses and brands looking to procure a wide spectrum of genuine handmade artisanal products produced in an ethical manner. Over time, this service became very useful for both ends. The rural artisan communities needed support, capacity building, design intervention, and quality upgradation to match buyer requirements. The buying houses, especially international conscious brands, could benefit from a 'one-stop window' to a vast array of artisanal products from across India, certified for genuineness and ethical production. This was a particularly important service sought by the buyers because they would not need to invest in researching for and finding materials, products, and producers from rural India to suit their needs.

Craftmark has actively engaged in negotiating for artisans' rights and protection of their economic interests. For example, Craftmark advocated for home-based traditional production processes *vis-a-vis* the standard factory compliances demanded by certain international buyers. It also pushed back against the notion of child labor being brought up as an ethical issue by international brands, educating the buyers about how India's traditional handicraft industry is essentially family-based, where skills are transmitted through generations. Culturally, children of artisans naturally learn the craft by doing, and engage in production activity alongside their family members; there is no exploitation in the process. Consumer education has been an important activity of the program to develop appreciation, recognition, and willingness to pay premium prices for high quality, unique, locally made handicraft products bearing distinctive cultural significance.

Craftmark remains a unique case study in India, addressing the challenges and advantages of the changing market scenario, patterns of cultural consumption, and emerging notions of sustainability impacting the handicraft value chain.

iTokri

iTokri[6] started as one of the first e-commerce brands of handmade artisanal products of India in 2011. It is a successful curated e-store with its base in Gwalior, in Madhya Pradesh. It showcases handicrafts, craftspeople who make those products, and non-profits who support artisan clusters. iTokri has a direct linkage to a network of 500+ artisan families, from whom they directly source products on a regular basis, impacting thousands of lives in the craft villages of India (Raja 2022).

During its inception, its founders realized that a mainstream corporate model of scaling up and growth would not work for the craft sector owing to its unique features across the diversity of crafts, maker communities, and regions. Hence, ideation of strategies will have to be done as per the gaps and challenges of particular crafts and the sector as a whole. A fundamental weakness of the craft product supply chain that the founders of iTokri identified and wished to address directly is that the actual craft artisans do not have the financial strength and the capacity to hold inventory after production and dispatch those products from their rural doorsteps as per orders. Instead of putting that responsibility on the artisans, iTokri decided to adopt a warehouse model and a buy-out approach for their business. Their understanding of the sector, and non-commission-based approach, was exactly what the artisans needed to be able to focus well on their own skills of craft and product making. Their assurance of a buy-out approach came with a strong commitment from the producers towards maintaining the high quality of the products. iTokri's model has tried to satisfy both producers and customers. They have created a one-stop shop for the diverse handicrafts of India for the buyers to choose from and buy. At the same time, they have consciously tried to provide flexibility to the artisans by making upfront payments to them upon procurement. This arrangement has enabled the craft producers to continue production, diversification, and innovation, thus strengthening their product base. Thus, by taking care of end-to-end logistics, iTokri is believed to have created a sustainable model for artisans and customers alike, directly addressing the challenges of a highly fragmented handicrafts sector in India.

Set up as a small business, iTokri has now expanded its reach across India both for sourcing as well as sales. iTokri buys craft products directly from the artisans, carries out professional product photography and curation on their e-commerce portal, holds the inventory, and couriers products on orders. Their branding strategy has always been to promote not only the product but also the makers and their traditions. Hence, artisan and craft details are also mentioned along with the product to inform buyers. The transparency in their business and moderate pricing are the factors that make them stand out as unique, and highly effective in this sector. They market a huge diversity of authentic Indian crafts, spanning different product segments and categories, making them accessible to all. iTokri has also ventured into making their own design collections which they send out to the artisans for production. These collections are exclusive to iTokri and are sold under their brand name.

iTokri enjoys a large, dedicated customer base, across India and overseas. Nearly 20% of their clients are from the UK, the US, and Canada. Their client segments include direct-to-consumer brands, retail, and e-commerce. Social media channels have created a large following of their brand, and word-of-mouth promotion has been one of the major factors driving sales. Their business is also managed by a completely local workforce from Gwalior, of whom 70% are women.

Conclusion

The scale and diversity of Indian handcrafts is quite overwhelming, and they are continuously evolving, impacted by modern markets, globalization, changing trends and fashion, exposure, interactions and exchanges with consumers, other businesses, fellow artisans, etc. In spite of such a huge cultural wealth of the nation, rural practitioners have remained marginalized, underprivileged, poor, and unrecognized, owing to a range of factors discussed in Chapter 1. This chapter highlights the immensely positive impact of conscious and visionary private sector entities which have worked in collaboration with rural entrepreneurs to uphold the creative industry of India in the international and national markets through changing times. Empowerment, value creation, and recognition of the actual producers of the handicrafts have been pivotal to their brand building and sustainability. The private sector ecosystem has played a vital role in transforming the dying, unknown, devalued rural handicraft skills of India into a vibrant, profitable, internationally acclaimed creative industry.

The case studies also contribute to the empirical knowledge related to concepts of 'entrepreneurship,' 'social entrepreneurship,' and 'cultural entrepreneurship' which have been active areas of academic research in the fields of business, and cultural and creative economy. The different models discussed here lie at the intersection of goals that may nominally seem at odds with each other: generating profit and managing business for cultural and creative goods, creating socio-cultural value for the creators of artistic and cultural goods, bringing about economic empowement of the creative producers themselves, and meeting the spiritual needs and building appreciation towards the previously devalued or unknown arts and cultural goods. In the academic body of work, although there is no single cohesive definition of social or cultural entrepreneurship, the various elements of these case studies feature in academic literature across disciplines.

Entrepreneurship has been defined as the process of generating business profits and creating economic value in the face of risk and through innovation. Social entrepreneurship is broadly viewed by some scholars as a business venture using innovation, generating revenue, and working with the core mission of doing social good, and the approach of applying the principles of entrepreneurship in the social sphere (Martin and Osberg 2007). The key characteristics of an entrepreneur, consistently emphasized by all, include being a risk-taker, an innovator, and attempting to create economic value or business profits through focused use of opportunities. In the case of a social entrepreneur, the existing definitions converge on the idea that the initiatives of the social entrepreneur are driven by the social objectives of benefiting society in some way and increasing social value by contributing to the well-being of the members of society. It has been observed by various researchers that profitability is consistent with social entrepreneurship, but the goal of profit generation is inextricably linked to generating social benefits and is the dominant motive of these ventures. Similar aspects are found to be driving forces for some of the entrepreneurship case studies presented in this chapter. However, our discussion would remain incomplete without reference to the definition or aspects of cultural entrepreneurship.

Cultural entrepreneurship is a relatively new concept in the academic field, the understanding of which has been explored by several scholars. The common converging ideas include a business enterprise based on the integration of two activity areas—culture and business (Madgerova and Kyurova 2019). The business activity is either undertaken by the creator or those who have the authority to organize the development, production, and marketing of cultural products of the producers, leading to economic, cultural, and social impact. The essential qualities of the cultural entrepreneur are to successfully utilize or generate the market for specific cultural products or services, as well as create innovative and new business models that lead to the market success of the business venture or the entrepreneurship. Therefore, a cultural entrepreneur operates in a specialized field of business wherein the entrepreneur not only understands business, revenue, and market dynamics but also has the knowledge and passion for applying creativity, design, and technology, constantly, to innovate the cultural good and products in response to the changing market tastes and consumer demands. Cultural entrepreneurs need to act in a constantly changing environment, owing to the fast-paced development of technology, and a globalized market and audiences whose tastes are constantly changing. Globalization has also resulted in unpredictable changes in the way the value chains of creative and cultural goods and services operate, transforming ways of production, presentation, innovation, and value creation. This is considered a unique business environment where uncertainties need to be accepted and social trust needs to be established. Thus, risk management and trust development are identified as central features in the establishment and development of cultural businesses (Banks et al., 2000). Entrepreneurship in culture and the arts has acquired an important economic and social role, especially in recent times, as it not only involves the creation of cultural goods of a tangible and intangible nature, promotion of their creators, and creation of cultural value, but also contributes to new job creation, social inclusion, micro-enterprise development generating grassroots entrepreneurship, and growth of national and regional economies.

As the cultural entrepreneurship models presented here deal with cultural producers living in rural India, who are relatively informal and are immensely diverse based on their ethnic backgrounds, cultural traditions, geographies, and use of local materials, it is imperative for the entrepreneurs to continuously innovate new strategies and business models to manage backend sustainably, introduce market novelties, as well as continuously add value and innovation to the cultural products and services for market success.

Notes

1 www.fabindia.com/ (accessed May 2023).
2 www.anokhi.com/ (accessed May 2023).
3 www.dastkar.org/ (accessed May 2023).
4 www.goodearth.in/ (accessed May 2023).
5 www.craftmark.org/ (accessed May 2023).
6 www.itokri.com/ (accessed May 2023).

References

Banks, Mark, Andy Lovatt, Justin O'connor, and Carlo Raffo. 2000. "Risk and Trust in the Cultural Industries." *Geoforum* 31 (4): 453–64.
Kapoor, Aekta. 2017. "The Heritage Chic." *Open Magazine*.
Konwar, Sunita Gupta. 2011. "Decoding Wellbeing-Oriented Business Model of Fabindia." *International Journal of Management Research* 43: 43–53.

Kumar, Adarsh. 2006. "Anokhi: In the Craft of Block-Printing." *Economic and Political Weekly* 41 (31): 3371–3374.

Madgerova, Raya, and Vyara Kyurova. 2019. "Specifics of Entrepreneurship in the Field of Cultural and Creative Industries." *Entrepreneurship* 7 (2): 103–23.

Martin, Roger L, and Sally Osberg. 2007. "Social Entrepreneurship: The Case for Definition." *Stanford Social Innovation Review* 5 (2): 29–39. https://doi.org/https://doi.org/10.48558/TSAV-FG11.

Raja, Vidya. 2022. "Couple Earn Rs 27 Crore While Helping 10,000 Artisans Take Their Work Across the World." www.thebetterindia.com/288184/couple-quits-job-itokri-empower-artisans-crores-profit-startup/; The Better India.

Tyabji, Laila. 2003. "Tradition and Transition: A Crafted Solution to Development." *The Future Is Handmade: The Survival and Innovation of Crafts (Special Issue of the Prince Claus Fund Journal)* 10a: 122–33.

7 Associated industries
Festivals and tourism

Festivals

Festivals are important spaces for accelerating the growth of creative and cultural industries. The practice of festivals has existed worldwide since the age-old times, to celebrate different aspects of life and society. Festivals include different communities, rituals, social customs, faith, cultural traditions, and a spirit of celebration, encompassing a complete cultural experience for a participant. Although festivals have been integral to local community lifestyles, rural and urban, it has also acquired a significant role in contributing towards local and national economies, building social cohesion and solidarity, and promoting cultural diversity and appreciation at local, national, and international levels. Hence, globally, national governments and local stakeholders have been using festivals for regional development and tourism.

India is well known as a land of festivals. Innumerable festivals happen all over the country celebrating local beliefs and faiths, various life events such as harvests, indigenous new years, and changes in seasons, worshipping indigenous and mainstream gods and goddesses, major religious events, as well as presentations of arts and crafts heritages of the country. All these occasions create scope for business transactions for various creative industries, such as food, decoration, gifting, cultural dresses, and ritualistic objects. The transactions take place within and outside the custodian communities leading to a multiplier effect boosting the local economies. Beyond maintaining traditions, festivals stimulate the creative economy by being distribution platforms for the associated creative industries. As the celebrations bring together different

DOI: 10.4324/9781003331476-7

communities, people, tourists, creative and cultural professionals, artisans and performers, and a large audience from different walks of life, a festival creates opportunities for the exchange of new ideas, forging new connections and partnerships, new investments, and new consumers. Moreover, there is an organic promotion of the festivals by tourists and visitors through word-of-mouth, articles, blogs, and social media, which strengthens the economy around the festival further.

The significance of festivals has increased far more in a globalized world where the homogenization of cultures and depletion of local cultural identities and traditions have become prevalent. Festivals as collective spaces of specific cultural activities and exchanges make them ideal platforms for the promotion of cultural diversity, recognition of indigenous cultures and creative industries, strengthening community identity and community organizing, especially of the marginalized cultures and people, and awareness creation about the specialities of a place. Overall, it is one of the most powerful avenues for uplifting creative industries and the economy.

There are many large festivals of India that are major revenue-generating events involving creative industries of different kinds. Some of the large-scale festivals driving creative businesses are Diwali (Festival of Lights), the local New Years (Bihu, Onam, Poila Boisakh, and others), Ganesh Chaturthi (Festival of Lord Ganesha, Hindu god of prosperity), and Durga Puja (Festival of Goddess Durga) which was inscribed in the UNESCO Representative List of the Intangible Cultural Heritage of Humanity in 2021. These festivals have become collective celebrations surpassing religious and social divides, leading to extensive mass participation. The state and national economic strategies particularly focus on these festivals for strengthening the state and national GDP. The following section presents a case study on the Durga Puja festival of West Bengal.

The Durga Puja festival of West Bengal

Durga Puja of Kolkata, in West Bengal, which got enlisted in the UNESCO list of Intangible Cultural Heritage in 2021, is one of the largest public arts festivals in the world encompassing an entire creative economy. This section describes the Durga Puja festival of West Bengal, but it is important to note that this festival is celebrated by the Bengali communities settled across the world. What started as a religious festival initiated by some rich families of Bengal has eventually transformed into a global cultural festival having an economic footprint of INR 32.4 crore (USD 4.53 billion) annually. Despite being just a weeklong festival, Durga Puja accounts for 2.58% of West Bengal's GDP. Of these, retail (INR 27,364 crore), food and beverage (INR 2,854 crore), and installation, arts and decoration (INR 860 crore) have the largest share (Virani et al. 2021).

Durga Puja is a five-day festival of worship and rituals for venerating Hindu gods and goddesses. It is a celebration around the homecoming of Goddess Durga (the divine power who defeated evil) and her four children who represent the divine powers of wealth, creativity, courage, and prosperity. Stories and folklores related to this festival have their origins in Hindu mythologies and epics, making it a deep-rooted and old cultural tradition of Bengal. According to historical accounts, the first big Durga Puja can be traced back to the late 1500s, believed to have been initiated by the local landlords of Bengal. Other records mention that a king in the Nadia district of Bengal organized the first Durga Puja. The practice of hosting Durga Puja festivals became popular in the households of Hindu kings and rich landlords, making it a tradition of the undivided Bengal region. The household-based festivals eventually evolved into local community festivals. The first community Durga Puja in Bengal was organized in 1790 by collecting contributions from the neighborhood. The first community-centered Durga Puja festival in Kolkata was organized in 1832 by the then king of Cossimbazar, who earlier used to host the puja in his ancestral home in Murshidabad from 1824 to 1831.

Over time, the fervor around the festival increased and the first large people's festival with full public contribution, management, and participation was organized by a community collective in Baghbazar of north Kolkata in 1910. Soon it turned into the biggest annual Autumn festival of Bengal and became the epitome of diverse creative expressions and industries.

Today, the festival is organized across urban and rural Bengal and involves several highly skilled artists, sculptors, and craftspersons. There are specialized communities that sculpt idols of the gods and goddesses out of hay and clay, paint them with colors, and adorn them with gorgeous attires, jewelry, and other accessories including various ritualistic items. The entire cohort of five deities are taken to different neighborhood festivals as per their orders and are placed in beautifully ornamented 'pandals' or temporary tent-like structures primarily made of bamboo, wood, and fabric and decorated with a multitude of innovative arts and crafts, such as paintings, terracotta, bamboo and cane items, wood carving, shola work, pottery, and paper craft. A different set of artisans specialize in pandal making and pandal decorations, which have become thematic in style exploring new themes, materials, art, and design every year. This has led to a creative burst with the aim of making every neighborhood festival unique and distinctive. For the festival, a specific type of percussion instrument called *dhaak* has ritualistic significance. The instrumentalists, called *dhaakis*, are another specialized community, thousands of whom get engaged in the numerous pandals of Bengal during the five days of the puja. Another important traditional industry linked to the festival is the decorative lighting industry which is engaged in lighting up the puja neighborhoods and the pandals with curated thematic lighting. The priests who carry out the puja are engaged professionally in all the puja locations, including both neighborhood pujas and those in the households.

The festival, which turns Kolkata city and various parts of Bengal, into a walk-in open-air curated art space, is also an occasion for celebrating literature, traditional food, traditional attires, etc. Hence, several other creative industries such as cuisine, fashion, textiles, footwear, cosmetics, as well as literature and publishing, tourism, and entertainment businesses peak around this period. Special festival editions of popular Bengali magazines are published, which are customarily bought by every Bengali household. The neighborhoods are filled with traditional and special food stalls as part of the festivities. People dressed in new clothes are found thronging in the pandals at any time of the day and night. Women adorn themselves in the best of the traditional textiles bought especially for the festival. Gifting of new apparel to relatives and friends is a social custom, generating business for retail brands and local shops alike.

With the expansion of the number of community pujas and the associated business sizes, corporates started engaging with Awards for best pandals, lighting, idols, and many other categories. This has boosted the creativity and scale of the puja organizing committees, the artisans and the artists, making the festival one of the biggest creative industries in India and the world. With more than 40,000 community pujas across the state, including 3,000 in Kolkata, the festival creates a range of economic activities every year which start about 3–4 months prior to the festival and peak around the five days of the festival. According to Virani et al. (2021), the retail segment witnesses a 100% increase in monthly sales value in West Bengal during this period, primarily driven by an increase in purchasing power and elevated spending sentiment. In 2019, registered pujas accounted for INR 700 crore, while unregistered pujas accounted for INR 160 crore. The typical budget of a 'Super-Mega' registered puja in Kolkata was INR 2.5 crore. Other sectors with a large share in the puja creative economy included advertisements (INR 504 crore), sponsorships (INR 318 crore), idol making (INR 260–280 crore), literature and publishing (INR 260–270 crore), and lighting and illumination (INR 205 crore).

Tourism

Tourism is one of the most profitable industries across the world leading to job creation, economic growth, and infrastructure development. Being one of the oldest civilizations of the world, India is a major global tourist destination, offering not only a diverse natural landscape but also an extremely rich cultural landscape. Tourism is India's third largest sector, employing about ten million people directly or indirectly. Its revenue stood at USD 247.3 billion in 2018, contributing 9.2% of the entire economy. The sector is also one of the largest earners of foreign exchange (IBEF 2022).

The UNWTO defined cultural tourists, at its 2017 General Assembly, as those who visit a tourism destination to 'learn, discover, experience and consume the tangible and intangible cultural attractions/products,' which 'relate to a set of distinctive material, intellectual, spiritual and emotional features of a society that encompasses arts and architecture, historical and cultural heritage, culinary heritage, literature, music, creative industries and the living cultures with their lifestyles, value systems, beliefs and traditions.'[1]. It estimates that more than 39% of all international tourism arrivals are cultural tourists (World Tourism Organization 2018; Richards 2018).

Balancing the interests of the tourism industry with proper heritage management, especially in the context of tangible heritage, is a well-known problem (McKercher and Du Cros 2002). Another critical aspect of tourism is sustainability in terms of reduced ecological footprint and improved community well-being. According to the World Charter for Sustainable Tourism 2015,[2] a tourism ecosystem should support responsible and sustainable tourism, including responsible tourist behavior, environment friendly tourism offerings, and improved community custodianship of their traditions, and natural and cultural assets. As a follow-up to this approach, policies for sustainable community tourism now take into consideration guidelines for equitable partnerships, community as a driver of cultural tourism, and ethical aspects of cultural consumption.

Cultural tourism in India

Cultural tourism is a growing tourism vertical in India, owing to its vast tangible and intangible cultural heritage spread across the subcontinent. India has 40 UNESCO World Heritage Sites, out of which 32 are 'cultural tourism' sites.

The Indian government has invested in policies and programs to support cultural tourism. The 'Incredible India!' campaign launched in 2002 has been extremely successful in bringing to the fore unknown or lesser-known cultural attractions of India to the world (Kant 2009). These include the creative industries of food, festivals, music, performing arts, architecture, traditional customs, and languages. A large private sector works towards inbound tourism boosting business, promotion, and incomes of the tourism stakeholders and beneficiaries.

Most of the unique intangible cultural heritages in India are local, rural, embedded into the way of life of the traditional rural communities, and call for safeguarding, innovation, community empowerment, and capacity building for managing responsible tourism, along with necessary infrastructural facilities. Thus rural cultural tourism is a niche area within the cultural tourism segment that entails community skill development, pro-poor growth, and inclusive development. To support such activities, stakeholders and change agents in the tourism, culture, and creative entrepreneurship sectors are working together to create responsible local cultural tourism destinations. Benefiting the local communities and protecting their local biodiversity and habitats are the major areas of focus for a successful and sustainable way forward.

To address the potential gaps in cultural tourism development, new markets and new brands are being developed by different stakeholders to promote village destinations rich in traditional art and crafts. Culture provides distinctiveness to a destination that is unique to the locality. Conversely, tourism provides resources to enhance culture and create income which can support and strengthen cultural heritage, cultural production, and creativity (OECD 2008).

There are many examples across India that demonstrate successful models of economic and social empowerment of local communities through cultural tourism initiatives. It has been found that cultural assets are not only sources of economic growth but also social inclusion, equity, and cultural empowerment. In totality, it leads to the collective well-being of the communities engaged in these initiatives.

Conclusion

As discussed in the preface, the 'creative economy' is a vast sector that includes multiple specialized creative industries and occupations, as well as associated industries. This chapter has briefly discussed two interesting associated industries of festivals and cultural tourism, based on local cultural resources, which significantly contribute to economic development, as well as create spaces for the amalgamation of cultural and social elements that energize a community and a place in a holistic way. There is much more to these industries that can be studied in greater detail.

The COVID-19 pandemic hit the tourism sector like never before, making local tourism an essential asset for the industry. The need of the hour is a responsible tourism sector founded on sustainability and built on local resources and opportunities. This is one of the central elements of UNWTO's Global Guidelines to Restart Tourism (UNWTO 2020). Opportunities of local cultural tourism leading to local community empowerment and strengthening of creative practitioners' engagement in local festivals are becoming the most effective 'new normal' strategies that focus on local culture, domestic visitors, and the benefits both to local communities as well stakeholders of this industry.

Notes

1 www.unwto.org/tourism-and-culture
2 www.gdrc.org/uem/eco-tour/charter.html

References

IBEF. 2022. "India's Tourism Sector on the Rise." www.ibef.org/blogs/india-s-tourism-sector-on-the-rise; India Brand Equity Foundation (IBEF).

Kant, Amitabh. 2009. *Branding India: An Incredible Story.* HarperCollins, Noida, India.

McKercher, Bob., and Hilary Du Cros. 2002. *Cultural Tourism: The Partnership Between Tourism and Cultural Heritage Management.* Haworth Hospitality Press, New York.

OECD. 2008. *The Impact of Culture on Tourism.* Organisation for Economic Co-Operation and Development (OECD). https://doi.org/https://doi.org/10.1787/9789264040731-en.

Richards, Greg. 2018. "Cultural Tourism: A Review of Recent Research and Trends." *Journal of Hospitality and Tourism Management* 36: 12–21. https://doi.org/https://doi.org/10.1016/j.jhtm.2018.03.005.

UNWTO. 2020. "UNWTO Launches Global Guidelines to Restart Tourism." UNWTO. www.unwto.org/news/unwto-launches-global-guidelines-to-restart-tourism.

Virani, Tarek, Morag Shiach, Joy Sen, Leandro Valiati, and Tanima Bhattacharya. 2021. "Mapping the Creative Economy Around Durga Puja." British Council India. www.britishcouncil.in/programmes/arts/Mapping-Creative-Economy-around-DurgaPuja.

World Tourism Organization. 2018. "Tourism and Culture Synergies." UNWTO, Madrid. https://doi.org/https://doi.org/10.18111/9789284418978.

8 Discussion

In the final chapter of this book, we summarize the key learnings from the case studies presented in earlier chapters and reflect upon the past and present conditions that enable and challenge the survival, growth, and health of creative economies based on traditional cultural heritage in a globalized world. The intention is not to provide or prescribe any particular solution or judge any specific model or approach. Rather, it is to reiterate the critical elements of traditional creative industries that are increasingly relevant in a fast-evolving technology-driven world, in relation to some theoretical and policy perspectives.

Cultural entrepreneurship

The history, evolution, and case studies in the preceding chapters clearly show that the Indian handicrafts industry has considerable cultural and social capital that can significantly benefit the creative economy in terms of sustainable livelihood and decent employment, gender equality, integrated rural development, socio-economic resilience, social solidarity, and environmental sustainability. The safeguarding and continuation of traditional skill-based creative industries can enhance the holistic advancement of the local communities towards economic, social, and emotional well-being. Rural creative communities who have practiced their craft forms for generations have an emotional connection with what they create. For them, the handcrafted products are not just sellable items but are essentially carriers of their tradition and culture. Hence, having a sustainable livelihood and full-time

DOI: 10.4324/9781003331476-8

work in what they love and excel in can positively contribute to the issues of poverty alleviation, the creation of dignified employment for all, and the growth of Indian handmade manufacturing industries, at the same time creating a unique position for the creative economy of India in the world market. The diversity and specialization of hand skills, along with indigenous knowledge of local materials, biodiversity resources, and the environment, are so rich and vast that they provide the country with a competitive edge over other creative economy players.

Historically, India has experienced multi-cultural exchanges that have enabled its creative industries to grow, diversify, and evolve through changing times. In the 1990s, economic liberalization opened up new markets and created a wide range of opportunities for creative entrepreneurship. Similarly, in today's globalized world, access to global creative initiatives, projects, industries, and traditions has been made easy for all. Artist-to-artist collaborations, creative exchanges, art and craft residencies, international festivals and exhibitions, and increased mobility of artists, artisans, designers, and entrepreneurs have boosted innovations, co-creations, new ideas, and amalgamation of creativity beyond borders, leading to exclusive and ingenious products and offerings benefitting both the producers and the markets. These multi-cultural partnerships have additionally boosted cultural solidarity, creating appreciation for indigenous art, cultures, and traditions from across the world, and generating awareness about various types of creative industries and communities, thus inspiring creative producers with new experiences. There is regenerative energy and a focus on creativity and culture as tools for sustainable and people-centric broad-based development, which need to be harnessed and channeled in a meaningful and rewarding way.

Although studies on creative and cultural entrepreneurship are relatively new to academic discourse, the topic has been well studied. A review of the literature on cultural entrepreneurship by Dobreva and Ivanov (2020) puts together multiple academic perspectives that are interesting. They note that the characteristics and motivations of cultural entrepreneurs differ from those of other entrepreneurs, because of their unique contextual and sectoral features, the particular nature of artistic work, and the associated cultural values that make the artistic work meaningful and complete. According to Klamer (2011), the creative process is the 'moral attribute' or cultural value, while the economic part is the means of realizing this value. Scott (2012) explains cultural entrepreneurship by dissecting it into three elements: creation of new cultural products, knowing how to access opportunities to establish an identity and a social trajectory, and finding ways to shape their creations without significant economic investments. What distinguishes them is their personal and emotional involvement in the creative process. Another characteristic of cultural entrepreneurs is the significance of place and social networking, where the place is the focal point for network building of cultural entrepreneurs and workers (Heebels and Van Aalst 2010; Lange 2011; Naudin 2017). This idea aligns with the essentiality of traditional spaces of handicrafts workers and entrepreneurs, which are the villages and their surrounding social, cultural, and environmental milieu. These spaces provide the right environment that inspires them to combine their talents and work with each other. The networking is a characteristic feature of cultural entrepreneurship, as they tend to spontaneously engage in cooperation, and collaboration (Coulson 2012). This adds to social solidarity and resilience.

The academic literature also reiterates the positive impact of cultural entrepreneurship in terms of generating new jobs and new models of work, creative production and economic growth, improved social cohesion, and a sense of belonging (Wilson and Stokes 2002). It emphasizes the vital characteristic of cultural entrepreneurs in combining individualistic values with collaborative work, in the process of engaging with the wider creative community. This behavior of cultural entrepreneurs explains the benefits and importance of cross-border artistic collaborations and festivals which enable them to grow and innovate in a constantly changing environment.

Empowerment of rural creative entrepreneurs

Although India's handicrafts-based creative economy is vibrant and successful in many ways, as evidenced by the selected case studies from different parts of the country, the broader business ecosystem has been fraught with common challenges. The most significant among those are the weak, and fragmented value chains of this sector, inadequate capacities of the rural producers in terms of organized production, access to quality raw materials, access to credit and finance, access to and knowledge of appropriate tools and technology, digital literacy, financial literacy, and regulatory compliance management. The ever-changing digital market environment has created several opportunities as well as challenges for the sector. Nowadays it is much easier to market products and carry out business from one's home, which can cater very well to village home-based craft production units. The creative producers cum entrepreneurs based in their own villages can take advantage of e-commerce and online selling platforms, build business relationships and clientele on mediums like WhatsApp and use social media for promotion, maximizing their profits through direct linkages with buyers. While these new information and communication technologies have brought audiences and consumers much closer to the cultural producers and entrepreneurs than before, effective use of digital technology for enhancing creative businesses in a holistic manner by these rural creative entrepreneurs remains a major challenge. The digital divide between rural creative producers and entrepreneurs and urban creative industries, led by people with formal education, has greatly reduced the business competitiveness of the former.

Lack of digital literacy and lack of organizational regulations for selling on online platforms are major bottlenecks for traditional producers based in the villages. A national drive towards digital empowerment of rural artisan communities and businesses, along with infrastructural enablement for digital facilities is essential to bridge the existing gap. The younger generations, who are internet and social media savvy, would be the ideal group of people to operationalize the digital facilities to enhance their creative businesses in the online market space.

Digital technology also poses a different kind of threat to the craft producer communities and enterprises selling authentic handmade products, as markets are being flooded with computer-generated traditional and modern designs imitating handicrafts and machine-made products at cheaper rates. The issue of cultural rights of traditional practitioner communities and the protection of IPR of traditional cultural and creative products have become the most challenging subjects in the modern digital era.

Even though there are excellent and successful private sector entities, those are dependent for the most part on passionate and entrepreneurial individuals who have invested in streamlining their own supply chains of rural traditional crafts and have co-created designs for the modern market with inputs of national and international fashion designers. The private sector has also grappled with quality issues and timely delivery when sourcing from village household units. In places where the craftspersons themselves have turned into successful entrepreneurs, it is often found that NGOs and non-profit organizations have carried out funded interventions to build the capacities of these entrepreneurs and their production groups in a systematic way, including long-term handholding support. In cases where the rural craftspersons have established their own business enterprises, they are often unable to deal with market environments either because they lack regulatory compliances, or they fail to manage timely production and fulfillment of orders. The absence of knowledge, information, access, and effective use of the right technology has hindered improved production efficiency, reduction of drudgery, improved process and product quality, and enablement of innovations. In the Indian context, larger issues that give rise to many of the other challenges are poverty and marginalization, and lack of inclusive approaches towards economic, social, cultural, and political equity.

Thought leaders have pointed out that although a broad-based growth driven by the private sector, and an overall climate fostering entrepreneurship, job creation, security, and benefit rights are crucial, it is essential to enable micro and small enterprises and empower the marginalized rural poor with necessary opportunities and investments (Narayan 2002). There are several government policies and schemes supporting micro and small enterprise development, including ones for the handicraft sector. The different schemes include training, health insurance, old age pension, development of common facility centers, technology inputs, and organization of exhibition cum sale nationally and internationally. However, the access, uptake, management, and monitoring of these schemes remain inadequate (AIACA 2017). Owing to the huge diversity and a very large base of artisans across the length and breadth of the country, India still does not have a national database of its traditional crafts industries, producers, or artisans, which makes it immensely difficult to plan effective national policies and programs.

The rate of labor force participation of women in India is low compared to the global average (Afridi, Dinkelman, and Mahajan 2018). An important cultural feature of the Indian handicraft industry that can help tap this unused potential is that it is largely women centric. In all handicraft traditions, women play a significant role in the production process. Being a home-based industry by nature, women engage in production work in between their household chores and child care. Often children also engage in the crafts with their parents, learning the skill in the process organically. Whereas in most weaving traditions, men are the primary weavers working on the handlooms, women are the ones who prepare the threads and carry out most of the pre-loom and post-loom work. Certain weaving traditions are dominated by women weavers too, such as those of Kota Doriya of Rajasthan and the back-strap loom weaving of north-east India. Among most of the hard material crafts and visual art, women play an equal role with men in design and production. In addition to the obvious economic benefits, a vibrant creative economy based on traditional handicrafts can therefore pave the way for women's empowerment and gender equity, leading to overall healthy families and villages. Although there are several case studies of successful rural women entrepreneurs and female leadership in the local creative industries, they still have a long way to go in terms of overcoming social stigma, cultural barriers, and marginalization. Women's participation in production activities, and their aptitude for business need to be professionalized and formally recognized within the business ecosystem.

These challenges need to be addressed through a concerted effort to build and strengthen the entire business ecosystem and holistic empowerment of the rural creative entrepreneurs and producers.

Academic literature treats 'empowerment' as multi-dimensional, similar to 'poverty,' as marginalized people require various types of assets and capabilities for their empowerment. Narayan (2005) lists four building blocks of empowerment—institutional climate, social and political structures, poor people's individual assets and capabilities, and their collective assets and capabilities. The institutional climate includes formal laws, regulations, and implementation processes driven by the state or other private and social institutions, as well as social norms of solidarity, exclusion, etc.

Individual assets include both physical and financial, and the right capabilities to enable individuals to use those assets effectively for their own well-being. Narayan also points out that social capabilities, including social solidarity, collective capacity to organize, collective sense of identity, and values related to collective well-being, are equally important for ensuring independent representation, voice, and overall life enhancement. This discourse aligns with the theoretical and pragmatic positions of the economists and social reformers discussed in Chapter 2.

Cultural entrepreneurship, ecological sustainability, and human well-being

In the modern globalized world, with different types of markets and buyers, there are requirements for adhering to constantly evolving norms and standards of authenticity, quality, fair practice, and environmental sustainability. The lack of standards and benchmarks for skills, quality, authenticity, social values, and environmental sustainability of production processes reduces the business competitiveness of rural craft entrepreneurs. In this regard, one may also argue that these benchmarks laid out by the western markets do not take into consideration the inherent sustainable qualities of the traditional Indian handicraft sector. A culture of social and environmental sustainability was ingrained in the traditional Indian lifestyles but has been fast diminishing. It needs to be understood that the integral link of Indian crafts to indigenous cultures and people's lifestyles makes it a unique creative business sector.

The grassroots case studies discussed in Chapters 4 and 5 demonstrate a range of sustainability components such as strengthening the local economy based on local resources, and building social capital and equity based on specialized skills of communities leading to collective working, production, and returns. These values are not new but have found fresh context as the world gravely suffers from unsustainable economic, social, and environmental conditions, highlighted particularly by the disruption of the economy during and after the COVID-19 pandemic. Echoing the same thoughts, a Peoples Sustainability Manifesto was launched as early as 2012, on the final day of the Rio+20 Summit.[1] The Manifesto was prepared through a consultative process engaging hundreds of civil society organizations who pledged to take action beyond the Rio+ 20 Summit, based on 14 sustainability treaties agreed upon, for transitioning into a sustainable future. The overarching demand from civil societies reflected in the Manifesto was equity within and across generations and between humans and nature, economic localization promoting sustainable lifestyles and livelihoods, respect towards cultural diversity, and collective response towards transitioning to a sustainable world order.

The culture of fast fashion in a globalized world, transitional consumer tastes, and market behavior influenced by a multitude of product choices have led to the devaluation of authentic artisanal handicrafts. This scenario calls for the safeguarding of traditional creative hand skills which is the foundation of the handicrafts-based creative industry of India. Many institutional efforts have been able to document, conserve, and revitalize dying handicraft practices. Government and non-government organizations have implemented certification schemes to protect the business interests of authentic handmade crafts. However, the proliferation of private businesses faking handicrafts products with machine-made goods has continued to compromise the business environment for the actual practitioner communities. Large and powerful conscious fashion and apparel brands, dedicated to promoting original handicrafts and their makers, have created a very large client base both nationally and internationally, but they still cater to the more conscious buyers who are interested in handmade cultural and heritage products and understand the value of the producer's skills and creativity. The mass market remains confused and unaware of the value and uniqueness of handmade products.

As an outcome of these conditions, two major shifts have happened in the traditional handicraft sector. The younger generations have been opting out of these industries, reducing and erasing the traditional skills. Those still practicing are replacing their traditional slow production methods with faster production processes using non-traditional and synthetic materials. For example, traditional folk painters who used to apply natural colors that they made themselves from local natural materials are replacing the natural colors with synthetic poster colors, thus losing a vital component of their tradition. Thus, the competitive values of the traditional crafts are getting lost, making the sector weak and impoverished. National policies need to support education and apprenticeship in crafts by investing in specialized, customized market-oriented training for youth, enabling them to pursue creative entrepreneurship with their traditional skills and crafts. Systematic learning mechanisms can help build and strengthen the already existing resources for a prosperous creative economy.

In recent years, there has been a shifting trend towards conscious markets dealing in sustainable production and consumption. In the context of global climate change and its adverse impacts, and issues of waste generation that is increasingly polluting the environment and damaging natural habitats, consumer brands have shifted towards sustainable production. Recycling and reusing are the key ideas driving responsible business and marketing globally. Large-scale factory-based producers are facing legally binding requirements to mandatorily adopt sustainable production practices and reduce environmental pollution. The larger the companies, the bigger the onus on them to implement sustainable and ethical methods at work. In contrast to the industrialized factory-based production systems, the traditional handicrafts industries are inherently sustainable based on hand skills, simple tools, and technology which are mostly hand-operated and require little or no electric power, using local and natural raw materials, with local community-led production. The scale of operation is usually much smaller, and the production time is much longer. The production systems are based on age-old knowledge and skills passed on through generations and are valued within the practitioner communities as their traditional heritage signifying ancestral links to their forefathers. The shift towards sustainable fashion in today's modern markets is encouraging slow production, use of natural materials without depleting the environment, creating chemical-free products, investing in timeless designs, using recyclable materials, and promoting the longevity of products rather than the 'use and throw' culture. This change in the consumer ethos has brought a renewed focus on handmade traditional products which India excels in. Indian handicrafts need consumer campaigns to educate consumers on the sustainable aspects of the Indian handicraft industry along with the values of community traditions and culture.

Although this paradigm shift is increasingly visible today in the tangible market and societal behavior, the approach itself has been addressed and discussed much earlier through various theoretical and policy frameworks including those of the United Nations. A particularly interesting and pertinent framework is that of Kothari (2014), which considers direct democracy, local and bio-regional economies, cultural diversity, people's well-being, and ecological resilience as the core indicators of development. As an alternative to the current economic engine, which focuses on global economic growth leading to the depletion of bio-cultural diversity and increased poverty, exploitation, and marginalization, he suggests a closer study of the different grassroots initiatives in India that have been trying to transform exploitative economic and social structures. Rather than a model of shared well-being led by the state or the market, he suggests putting collectives and communities at the center of achieving this goal. He emphasizes the need for greater localization where production and consumption patterns are embedded within the community. Economic democracy, according to him, should be harmonized with environmental sustainability, as the local communities living closest to natural resources have the greatest stake in them and the indigenous knowledge to manage them, in terms of efficient resource use, small-scale manufacture, and recycling efforts. The force of globalization can positively contribute to this process in terms of fostering a global community of humanity and reinforcing local sustainability efforts and good practices. These very ideas are reflected in the values and learning that the case studies in this book bring to the table for mobilizing and inspiring change towards a sustainable future.

Policy ecosystem

As cultural entrepreneurship is changing the face of the economy in recent times and is being considered as part of the main economic sectors, there is a need for supportive public policies for its growth. Without an enabling and professional ecosystem, its potential cannot be fully realized. Forging effective multi-stakeholder and public-private partnerships and collaborations, as well as making national programs and policies more flexible to accommodate R&D and innovations, are essential. Schulte-Holthaus (2018) highlights the need for highly skilled workers, better access to finance, promotion of cultural and creative industries to reach high export levels, and awareness and protection of intellectual property rights.

Internationally, as well as in India, cultural entrepreneurship and knowledge-based creative economy are gaining momentum through new policy thinking, and strategies for establishing them as strong economic sectors. Most recently, relevant policy agendas have emerged from the G20, an intergovernmental forum comprising 19 countries and the European Union, working to address major issues related to the global economy, such as international financial stability, climate change mitigation, and sustainable development. Some of the key outcome documents of the 18th G20 summit (September 2023) include the issues of culture, youth, women empowerment, principles of sustainability and resilience, circular economy, social protection, tourism, accelerating sustainable development goals, inclusive growth, digital economy, environment, and climate sustainability. As the G20 Culture Working Group[2] notes in its background paper, the creative economy is increasingly being considered an untapped resource for economic growth and employment of youth, building human capital, and promoting ecological sustainability. The creative economy in India is organically inclusive of social divisions and provides a strong cultural identity to the otherwise marginalized people that can contribute to their holistic empowerment. The sector is dynamic and diverse in terms of creative and cultural goods, services, and the inherent process of continuous innovation by the practitioners. Culture being a cross-cutting resource for other economic sectors as well, the growth of the creative economy drives new investments and skilling in other economic sectors, thereby boosting the national economy.

To support the growth of the creative economy, the Culture Working Group under India's G20 presidency plans to forge international cooperation among G20 members, to build upon the opportunities in the global creative economy to (i) strengthen the status and socio-economic rights of artists, craftspeople and cultural professionals, including in the informal sector, while enabling professionalization and skills development (ii) strengthen public policies framing the creative sector, including by supporting whole-of-government approaches and inter-ministerial collaboration (iii) advance the adaptation of the creative economy to the digital transformation, notably as regards fair remuneration, equitable access or the protection of online cultural and linguistic diversity, including by supporting a more inclusive global flow of cultural goods and service (iv) support the monitoring and measurement of the contribution of the creative economy to economic growth and sustainable development.

Way forward

The transformative power of culture for sustainable development has been realized and championed by various organizations, cultural professionals, and civil societies from across the world. Initiatives on achieving the sustainable development goals (SDGs) have demonstrated the transversal role of culture in poverty alleviation, social inclusion, resilience, gender equity, biodiversity conservation, sustainable consumption, and production, etc. Culture fosters innovation and creativity and provides a source of identity to the unique and culturally diverse communities in a globalized world.

Its universal contribution to holistic human development has also been reaffirmed through different global deliberations and exchanges in the post-pandemic era. However, culture as an essential component of human development still remains peripheral to mainstream development issues and dialogues. Public policies and investments towards culture as a resource for sustainable development have been insignificant in spite of long-term advocacy by the concerned stakeholders. Hence, at Mondiacult 2022, which was the largest conference devoted to culture convened by UNESCO in the last 40 years, 150 State Parties, along with a large international community of Civil Society Organizations (CSOs) and stakeholders, adopted a declaration affirming Culture to be a 'Global Public Good.' This declaration repositions culture as a universal asset not only for economic growth but also for the well-being of humankind and the protection of human rights, beyond private interests and gains.

The need of the hour is to collectively review the current situation of creative and cultural industries, international and local policies, and collectively advocate for including creative industries at the heart of integrative national and regional development policies. The process needs to be undertaken by multiple stakeholders beyond the silos of culture and welfare policies and administrations of governments, UN administrative bodies, and cultural professions; it must involve the allied sectors of Tourism, Education, IPR, and Technology, with support from academic fields such as Sociology, Economics, Statistics, Environmental Science, Public Policy, Law, and Business Management. More partnerships are necessary to strengthen and promote outcome-oriented integrative programs to mainstream the creative industry sector in economic and developmental plans. A critical task is to effectively measure the impact of creative industries on sustainable development, especially in the light of post-pandemic societal shifts in economic, social, cultural, and environmental priorities. Evidence-based case studies and best practices need to be promoted beyond borders for learning, adaptation from existing successful models, and for drawing investments towards supporting equitable ecosystem services for strengthening a vibrant and fair creative economy.

Notes

1 https://sustainabilitytreaties.org/ (accessed May 2023).
2 www.unesco.org/en/sustainable-development/culture/g20/creativeeconomy (accessed September 2023).

References

Afridi, Farzana, Taryn Dinkelman, and Kanika Mahajan. 2018. "Why Are Fewer Married Women Joining the Work Force in Rural India? A Decomposition Analysis over Two Decades." *Journal of Population Economics* 31: 783–818.

AIACA. 2017. "National Handicrafts Policy Report." AIACA.

Coulson, Susan. 2012. "Collaborating in a Competitive World: Musicians' Working Lives and Understandings of Entrepreneurship." *Work, Employment and Society* 26 (2): 246–61.

Dobreva, Nevena, and Stanislav Hristov Ivanov. 2020. "Cultural Entrepreneurship: A Review of the Literature." *Tourism & Management Studies* 16 (4): 23–34.

Heebels, Barbara, and Irina Van Aalst. 2010. "Creative Clusters in Berlin: Entrepreneurship and the Quality of Place in Prenzlauer Berg and Kreuzberg." *Geografiska Annaler: Series B, Human Geography* 92 (4): 347–63.

Klamer, Arjo. 2011. "Cultural Entrepreneurship." *The Review of Austrian Economics* 24 (2): 141–56.

Kothari, Ashish. 2014. "Radical Ecological Democracy: A Path Forward for India and Beyond." *Development* 57 (1): 36–45.

Lange, Bastian. 2011. "Professionalization in Space: Social-Spatial Strategies of Culturepreneurs in Berlin." *Entrepreneurship and Regional Development* 23 (3–4): 259–79.

Narayan, Deepa. 2002. *Empowerment and Poverty Reduction: A Sourcebook*. World Bank, Washington, DC.

———. 2005. "Conceptual Framework and Methodological Challenges." In *Measuring Empowerment: Cross-Disciplinary Perspectives*, edited by Deepa Narayan, 3–38. World Bank Publications, Washington, DC.

Naudin, Annette. 2017. *Cultural Entrepreneurship: The Cultural Worker's Experience of Entrepreneurship*. Routledge, London, New York.

Schulte-Holthaus, Stefan. 2018. "Entrepreneurship in the Creative Industries" In *Entrepreneurship in Culture and Creative Industries: Perspectives from Companies and Regions*, edited by Elisa Innerhofer, Harald Pechlaner, and Elena Borin. Springer, 99–154.

Scott, Michael. 2012. "Cultural Entrepreneurs, Cultural Entrepreneurship: Music Producers Mobilising and Converting Bourdieu's Alternative Capitals." *Poetics* 40 (3): 237–55.

Wilson, Nicholas, and David Stokes. 2002. "Cultural Entrepreneurs and Creating Exchange." *Journal of Research in Marketing and Entrepreneurship* 4 (1): 37–52.

Index